Field Manual
No. 3-34

*FM 3-34 (FM 3-34)

Headquarters
Department of the Army
Washington, DC, 2 April 2014

Engineer Operations

Contents

		Page
	PREFACE	iii
	INTRODUCTION	iv
Chapter 1	ENGINEER REGIMENT	1-1
	Engineer Disciplines	1-1
	Engineer Organization	1-3
	Operating-Force Engineers	1-4
	Engineer Force Tailoring	1-12
	United States Army Corps of Engineers	1-13
	Unified Action Partners	1-18
Chapter 2	ENGINEER SUPPORT TO UNIFIED LAND OPERATIONS	2-1
	Engineer Tasks	2-1
	Lines of Engineer Support	2-1
	Engineer Support to Warfighting Functions	2-8
	Tasks Supporting Decisive Action	2-15
Chapter 3	INTEGRATING ENGINEER SUPPORT	3-1
	Integrated Planning	3-1
	Staff Processes	3-4
	Other Tasks	3-18
	GLOSSARY	Glossary-1
	REFERENCES	References-1
	INDEX	Index-1

Distribution Restriction: Approved for public release; distribution is unlimited.

*This publication supersedes FM 3-34, 4 August 2011.

2 April 2014   FM 3-34   i

Figures

Introductory figure-1. Engineer framework.. iv
Figure 1-1. Engineer Regimental relationships... 1-3
Figure 1-2. BEB.. 1-6
Figure 1-3. Engineer companies 1 and 2.. 1-8
Figure 2-1. Engineer application of combat power 2-9
Figure 2-2. Notional engineer support to offensive tasks 2-16
Figure 2-3. Notional engineer support to defensive tasks 2-17
Figure 2-4. Notional engineer support to stability tasks... 2-18
Figure 2-5. Notional engineer support to DSCA tasks 2-21

Tables

Introductory table-1. Modified Army terms vi
Table 1-1. Operating-force engineers 1-5
Table 1-2. USACE division alignment 1-14
Table 3-1. Military decisionmaking process and engineer staff running estimate 3-10
Table 3-2. Engineer considerations in the military decisionmaking process 3-14

Preface

FM 3-34 is the Army doctrine publication that presents the overarching doctrinal guidance and direction for conducting engineer activities and shows how it contributes to decisive action. It provides a common framework and language for engineer support to operations and constitutes the doctrinal foundation for developing other fundamentals and tactics, techniques, and procedures detailed in subordinate doctrine manuals. This manual is a key integrating publication that links the doctrine for the Engineer Regiment with Army capstone doctrine and joint doctrine. It focuses on synchronizing and coordinating the diverse range of capabilities in the Engineer Regiment to support the Army and its mission successfully. FM 3-34 provides operational guidance for engineer commanders and trainers at all echelons and forms the foundation for United States (U.S.) Army Engineer School curricula.

To comprehend the doctrine contained in FM 3-34, leaders must first understand the elements of unified land operations, operational design, the elements of combat power, and the operations process as described in ADP 3-0 and addressed in ADP 2-0, ADP 3-37, ADP 4-0, ADP 5-0, ADP 6-0, and ADP 6-22. Readers must be familiar with ADP 3-07, ADP 3-28, and ADP 3-90. Leaders must understand how offensive, defensive, and stability or defense support of civil authorities (DSCA) operations complement each other. They must also understand the terms and symbols described in ADRP 1-02.

FM 3-34 applies to Army engineer forces. The principal audience for this manual is engineer commanders and staff officers, but all Army leaders will benefit from reading it. Trainers, educators, and combat developers throughout the Army also use this manual.

Commanders, staffs, and subordinates ensure that the decisions and actions comply with applicable U.S., international and, in some cases, host nation (HN) laws and regulations. Commanders ensure that Soldiers operate according to the law of war and the rules of engagement. (See FM 27-10 for additional information.)

Unless this publication states otherwise, masculine nouns and pronouns do not refer exclusively to men.

FM 3-34 uses joint terms where applicable. Selected joint and Army terms and definitions appear in the glossary and the text. Terms for which FM 3-34 is the proponent (the authority) are marked with an asterisk (*) in the glossary. Definitions for which FM 3-34 is the proponent publication are boldfaced in the text. For other definitions shown in the text, the term is italicized and the number of the proponent publication follows the definition.

FM 3-34 applies to the Active Army, Army National Guard/Army National Guard of the United States, and U.S. Army Reserve unless otherwise stated.

The proponent and preparing agency of FM 3-34 is the U.S. Army Engineer School. Send comments and recommendations on Department of the Army (DA) Form 2028 (*Recommended Changes to Publications and Blank Forms*) to Commandant, U.S. Army Engineer School, ATTN: ATZT-CDC, 14000 MSCoE Loop, Suite 270, Fort Leonard Wood, Missouri 65473-8929. Submit an electronic DA Form 2028 or comments and recommendations in the DA Form 2028 format by e-mail to <usarmy.leonardwood.mscoe.mbx.cdidcodddengdoc@mail.mil>.

Introduction

The Engineer Regiment exists to provide the freedom of action for land power by mitigating the effects of terrain. This manual explains how (not what) to think about exploiting the capabilities of the Engineer Regiment in support of military operations.

This version updates the engineer doctrinal framework (see introductory figure-1) that provides the intellectual underpinnings for the Engineer Regiment and better articulates its purpose and activities. It describes how engineers combine the skills and organizations of the three interrelated engineer disciplines (combat, general, and geospatial engineering) to provide support that helps ground force commanders—
- Assure mobility.
- Enhance protection.
- Enable force projection and logistics.
- Build partner capacity and develop infrastructure among populations and nations.

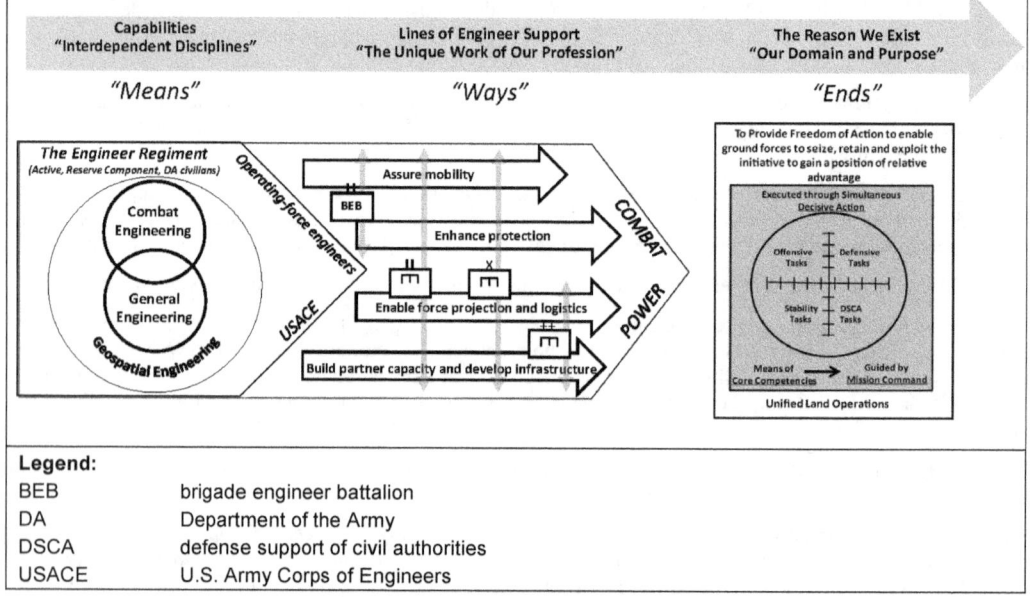

Introductory figure-1. Engineer framework

The update of this manual was driven by several factors, to include the—
- Establishment of the Doctrine 2015 Strategy by the 37th Chief of Staff of the Army in June 2011.
- Transition of the brigade engineer battalions (BEBs) into the brigade combat teams (BCTs).
- Adoption of the JP 3-34 definitions for combat, general, and geospatial engineering.
- Modification of the three engineer disciplines. The engineer disciplines remain interdependent on one another; however, the disciplines have been rearranged to reflect the relationship to one another. The engineer disciplines are also associated to the lines of engineer support. The geospatial engineering discipline is the foundation that supports the combat and general engineering disciplines and the lines of engineer support. The combat engineering discipline may support the lines of engineer support, but it tends to be more associated with the first two lines of engineer support. The

general engineering discipline supports the last three lines of engineer support most, but may support all four lines. (See introductory figure-1.)
- Introduction of means, ways, and ends. The disciplines are the means to which the Regiment applies its capabilities to achieve the ends. The ways are how the capabilities of the Regiment are applied to enable combat power. The ends provide freedom of action to enable ground forces to seize, retain, and exploit the initiative to enable unified land operations. Essentially, the Engineer Regiment consists of the three engineer disciplines that are found in the operating and generating force, which conducts multiple tasks along each of the lines of engineer support to enable combat power and ensure freedom of action.
- Modification of two of the four lines of engineer support—
  - *Enable force projection* was added to the enable the logistics line of engineer support. Our adversaries will attempt to compromise our ability to project combat power by the use of antiaccess and area denial methods by using hybrid threats. Early-entry forces must be capable of a rapid transition from deployment to employment. The enable force projection and logistics line of engineer support is intended to establish and maintain the infrastructure necessary for supporting early-entry and follow-on forces to sustain military operations.
  - *Build partner capacity* was added to develop the infrastructure line of engineer support. This line primarily, but not exclusively, consists of training and developing local leaders, engineers, and organizations, while conducting general engineering tasks with our partners so that they may effectively protect and govern citizens. It is the development of the intellectual thought required to build the institutions and processes to produce, manage, and regulate critical services. Developing infrastructure is the physical aspect of how engineers assist in enabling nations to effectively reconstruct and build critical infrastructure systems and essential services to protect and govern citizens.
- Realization that we are at an important crossroad, faced with limited personnel, materiel, and training resources that is compounded by a complex and uncertain future where hybrid threats will employ regular and irregular capabilities. The Regiment is faced with balancing reductions and meeting the need for reversibility. Our requirements are many, and our solutions must be optimized to support a wide range of military operations that can rapidly provide scalable capabilities to prevent, shape, and win the wars of the nation.

The doctrinal engineer foundations provided in this manual will support the actions and decisions of engineer commanders. But, like ADP 3-0, the manual is not meant to be a substitute for thought and initiative among engineer leaders. No matter how robust the doctrine or how advanced the new engineering capabilities and systems, it is the engineer Soldier who must understand the operational environment, recognize shortfalls, and adapt to the situation on the ground. It is the adaptable and professional engineer Soldiers and civilians of the Regiment who are most important to the future, and they must be able to successfully perform basic skills and accomplish the mission with or without the assistance of technology.

The operational environment will remain unpredictable with a wide range of threats from regular and irregular forces using conventional and unconventional capabilities, to include terrorist and criminal tactics. Our emphasis will continue to be in the Middle East and the Asia-Pacific regions. The Engineer Regiment has been reducing its Regular Army strength and restationing engineers from outside the continental United States to continental United States based locations over the past several years. This trend will continue. As a result, the reliance on Reserve Components will increase. Engineers must be able to support early-entry operations (including forcible entry where access is denied) to enable the seizure, establishment, and expansion of lodgments in an immature theater. We must be technically and tactically capable across the range of military operations (homeland and abroad) to support the Army strategic framework of preventing conflict, shaping the operational environment, and winning the wars of the nation.

Introduction

This edition of FM 3-34 covers the following information:
- Chapter 1 addresses the left side of the engineer framework, providing an overview of the Engineer Regiment and its organization and capabilities. It defines and highlights the interdependence of the engineer disciplines.
- Chapter 2 addresses the middle portion of the engineer framework, defining the four lines of engineer support and describing the relationships to the engineer disciplines, decisive action, and the warfighting functions.
- Chapter 3 describes how engineer support is integrated into the overall operation of the supported commander throughout the operations process. It describes engineer planning activities and considerations for preparing, executing, and continuously assessing engineer support.

Based on current doctrinal changes, certain terms for which FM 3-34 is the proponent have been modified. (See introductory table-1.) The glossary contains acronyms and defined terms.

Introductory table-1. Modified Army terms

Terms	Remarks
Combat engineering	Adopted joint definition.
Engineer work line	Definition modified.
Field force engineering	Definition modified.
General engineering	Adopted joint definition.
Geospatial engineering	Adopted joint definition.
Survivability operations	ATP 3-37.34 is the proponent.
Terrain reinforcement	Common English usage.
Legend: ATP Army techniques publication	

Chapter 1

Engineer Regiment

The Engineer Regiment is a military engineering profession within the Army profession that represents the Army engineering capabilities. The Engineer Regiment is the manifestation of this profession within the Army. It is composed of people—not just equipment, organizations, or technology—who serve with unique technical skills. These skills are grouped together into three engineer disciplines—combat, general, and geospatial engineering. It consists of Regular Army, Army National Guard, and U.S. Army Reserve engineer organizations; the U.S. Army Corps of Engineers (USACE); Department of Defense (DOD) civilians; and affiliated contractors and agencies in the civilian community. It has a diverse range of capabilities that are focused on providing the required engineer expertise and skills needed to support the combined arms team.

ENGINEER DISCIPLINES

1-1. The engineer disciplines are areas of expertise within the Engineer Regiment. Each discipline is focused on capabilities that support, or are supported by, the other disciplines. Within these disciplines are personnel and equipment that provide unique technical knowledge, services, and capabilities that make engineers a member of the Army profession.

1-2. Ground forces conduct operations on, in, or above the terrain. The ground forces are affected by the terrain, and they often affect it. Engineer operations are unique because, whatever the intended purpose, engineer operations are directly aimed at affecting the terrain or at improving the understanding of the terrain. In this context, terrain includes natural and man-made terrain features (obstacles, roads, trails, bridges, airfields, ports, base camps). As a result, terrain is central to the three engineer disciplines. Combat and general engineering are focused on affecting the terrain, while geospatial engineering is focused on improving the understanding of the terrain.

1-3. Regardless of the disciplines, engineers must be prepared to conduct missions in close combat. Combat engineering is the only discipline that is trained and equipped to support movement and maneuver while in close combat. The general and geospatial engineering disciplines are not organized to move within combined arms formations or to apply fire and maneuver. The general and geospatial engineering disciplines have small arms and a limited number of crew-served weapons that are capable of engaging in close combat with fire and movement, primarily in a defensive role.

COMBAT ENGINEERING

1-4. *Combat engineering* is the engineering capabilities and activities that closely support the maneuver of land combat forces consisting of three types: mobility, countermobility, and survivability (JP 3-34). This engineer discipline focuses on affecting terrain while in close support to maneuver. Combat engineering is integral to the ability of combined arms units to maneuver. Combat engineers enhance force momentum by shaping the physical environment to make the most efficient use of the space and time necessary to generate mass and speed while denying the enemy maneuver. By enhancing the unit ability to maneuver, combat engineers accelerate the concentration of combat power, increasing the velocity and tempo of the force to exploit critical enemy vulnerabilities. By reinforcing the natural restrictions of the physical environment, combat engineers limit the enemy ability to generate tempo and velocity. These limitations increase enemy reaction time and degrade its will to fight.

Chapter 1

GENERAL ENGINEERING

1-5. *General engineering* is the engineering capabilities and activities, other than combat engineering, that modify, maintain, or protect the physical environment (JP 3-34). This engineer discipline is primarily focused on providing construction support. It is the most diverse of the three engineer disciplines and is typically the largest percentage of engineer support that is provided to an operation, except in offensive and defensive operations at the tactical level when combat engineering will typically be predominant. It occurs throughout the area of operations, at all levels of war, and during every type of military operation. It may include the employment of all military occupational specialties within the Engineer Regiment. See FM 3-34.400 for additional information on general engineering operations.

1-6. General engineering is primarily focused on construction support. Tasks most frequently performed under general engineering include—
- Restoring damaged areas.
- Constructing and maintaining lines of communication (LOCs).
- Establishing base camps.
- Repairing and restoring infrastructure.
- Providing environmental assessments.
- Providing master facility and design support.
- Developing and maintaining facilities.
- Providing power generation and distribution.

GEOSPATIAL ENGINEERING

1-7. *Geospatial engineering* is the engineering capabilities and activities that contribute to a clear understanding of the physical environment by providing geospatial information and services to commanders and staffs (JP 3-34). Geospatial engineers generate geospatial products and provide services to enable informed running estimates and decisionmaking. It is the art and science of applying geospatial information to enable an understanding of the physical environment as it affects terrain for military operations. The art is to understand mission variables; apply the relevant geospatial information; and describe the military significance of the terrain and the other spatial and temporal aspects of the operational environment to the commander. The science is the exploitation of geospatial information and services, producing spatially accurate products for measurements, mapping, visualization, and modeling in support of the six warfighting functions. It includes the application of the Army geospatial enterprise (a distributed database) and a supporting infrastructure that is based on a common suite of interoperable software. The Army geospatial enterprise allows geospatial data to be collected, stored, fused, analyzed, and disseminated horizontally and vertically to provide the geospatial foundation for the common operational picture. See ATTP 3-34.80 and JP 2-03 for additional information on geospatial engineering.

1-8. Geospatial engineers provide the following support, from the Army service component command (ASCC) to BCT levels:
- Terrain analysis and other tactical decision aids that support the operations process.
- Terrain visualization, to include three-dimensional terrain mapping and fly-through representation.
- Nonstandard, tailored map products, to include cross-country mobility, view shed, zone of entry, and hydrology.
- Geospatial foundation data (maintaining, updating, managing, and disseminating) for the common operational picture.
- Theater geospatial database (maintaining, updating, and managing).

ENGINEER ORGANIZATION

1-9. The Army organizes Soldiers and equipment into a variety of organizations, each with particular capabilities. Engineer units are organized based on the engineer disciplines. The Engineer Regiment is composed of organizations that are represented in the operating and generating forces. These organizations operate concurrently with one another to support combatant commanders (CCDRs) and unified action partners. The Engineer Regiment Active Army organizations include USACE and Army military engineer units within the combatant commands and Army commands. Approximately three-fourths of Army military engineer units are in the Reserve Component. The Reserve Component provides two theater engineer command (TEC) headquarters, including a wide range of specialized capabilities in its Army National Guard and U.S. Army Reserve components. At the center of this is the Office of the Chief of Engineers. The Chief of Engineers integrates capabilities and supports the planning, preparing, executing, and assessing of joint operations. The Regiment is experienced at providing interagency support and in leveraging nonmilitary and nongovernmental engineer assets to support mission accomplishment.

1-10. The engineer branch includes the human resource managers in the Human Resources Command and the engineer branch proponent under the U.S. Army Training and Doctrine Command. Together, these organizations generate and manage engineer Soldiers. The engineer branch proponent is the U.S. Army Engineer School. It trains, educates, certifies, and manages engineer Soldiers.

1-11. The U.S. Army Engineer School provides specialized unit and individual training, including the Joint Engineer Operations Course, Route Reconnaissance and Clearance Course, Explosive Ordnance Clearance Agent Course, Search Advisor Course, Mine Detection Dog Course, and Sapper Leader Course. The engineer branch works closely with USACE to leverage a vast pool of additional technical engineer expertise provided by DOD civilians, affiliated contractors, and agencies within the civilian community. Technical support is available directly in support of the engineer staff and forces through the USACE reachback operations center. The Counter Explosive Hazards Center coordinates doctrine, organization, training, materiel, leadership and education, personnel, and facilities (DOTMLPF) policy solutions and integration to counter explosive hazards. Figure 1-1 shows how the Engineer Regiment is represented.

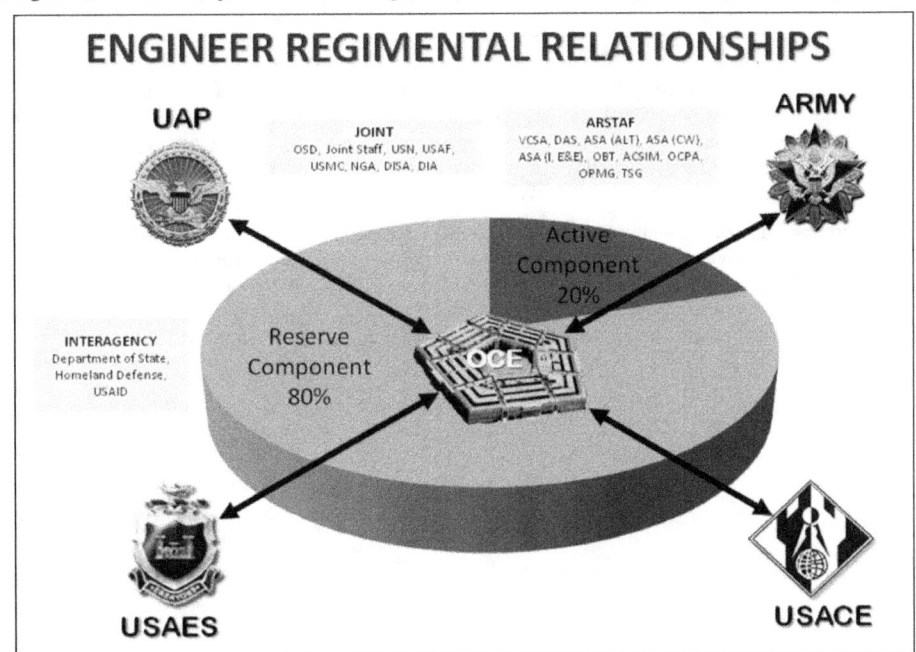

Figure 1-1. Engineer Regimental relationships

Legend:			
ACSIM	Assistant Chief of Staff for Installation Management	OCE	Office of the Chief of Engineers
ARSTAF	Army staff	OCPA	Office of the Chief of Public Affairs
ASA (ALT)	Assistant Secretary of the Army for Acquisition	OPMG	Office of the Provost Marshal General
		OSD	Office of the Secretary of Defense
		TSG	theater support group
ASA (CW)	Assistant Secretary of the Army (Civil Works)	UAP	unified action partner
		USACE	U.S. Army Corps of Engineers
ASA (I, E&E)	Assistant Secretary of the Army (Installations, Energy, and Environment)	USAES	U.S. Army Engineer School
		USAF	U.S. Air Force
DAS	Department of Administrative Services	USAID	U.S. Agency for International Development
DIA	Defense Intelligence Agency		
DISA	Defense Information Systems Agency	USMC	U.S. Marine Corps
NGA	National Geospatial-Intelligence Agency	USN	U.S. Navy
OBT	Office of Business Transformation	VCSA	Vice Chief of Staff of the Army

Figure 1-1. Engineer Regimental relationships (continued)

1-12. The Office of the Chief of Engineers is a staff element assigned to the Army staff to assist the Chief of Engineers in advising the Chief of Staff of the Army and the Army staff. The Chief of Engineers leads the Engineer Regiment and serves in three distinct roles—the chief of the Engineer Branch, the commander of USACE, and the staff officer advising the Chief of Staff of the Army on engineering matters and force capabilities. The chief is assisted in these roles by Headquarters, USACE; Commandant, U.S. Army Engineer School; and the Office of the Chief of Engineers. The Chief of Engineers is the senior engineer leader for decisions that affect the conduct of operations within the Regiment. As the head of delegation to the North Atlantic Treaty Organization (NATO) Senior Joint Engineer Conference, the Chief of Engineers is a key leader involved with interoperability of unified action partners. The Chief of Engineers is a member of the Joint Operational Engineering Board, which is the voice of the Regiment and makes decisions to provide the engineering capabilities that make engineers members of the Army profession.

OPERATING-FORCE ENGINEERS

1-13. Engineers in the operating force operate at the strategic, operational, and tactical levels across the range of military operations. Units are organized in a scalable, adaptable manner to support combat, general, and geospatial engineering requirements. Army engineer forces operate as integral members of the combined arms team during peace and war to provide a full range of engineering capabilities in conjunction with USACE. This section provides an overview of engineers in the operational force.

1-14. There are four complementary and interdependent categories of engineer units in the operating force, including USACE-provided technical engineering assets. The four categories include organic engineers (and staff elements) and three other categories held in an engineer force pool. The assets in the force pool reside at echelons above brigade (EAB) and exist to augment BCT engineers. The EABs consist of engineer headquarters units, baseline units, and specialized engineer units (see table 1-1).

Table 1-1. Operating-force engineers

Engineer Elements			Component		
			Regular Army	ARNG	USAR
Organic engineers		Brigade engineer battalion	X	X	
		Brigade combat team engineer company	X	X	
		Geospatial engineer team	X	X	
Force Pool	Engineer headquarters	Theater engineer command			X
		Engineer brigade headquarters	X	X	X
		Engineer battalion	X	X	X
	Baseline engineer units	Sapper company	X	X	X
		Mobility augmentation company	X	X	X
		Clearance company	X	X	X
		Engineer support company	X	X	X
		Horizontal construction company	X	X	X
		Vertical construction company	X	X	X
		Multirole bridge company	X	X	X
	Specialized engineer units	Survey and design team	X	X	X
		Concrete section		X	X
		Asphalt team		X	X
		Firefighting team	X	X	X
		Explosive hazard team or coordination cell		X	X
		Engineer squad (canine)	X		
		Diving team	X		
		Topographic company or platoon	X	X	
		Geospatial planning cell	X		
		Construction management team	X	X	X
		Engineer facility detachment		X	X
		Engineer utilities detachment		X	X
		Prime power company*	X		
		Technical rescue company		X	
		Well drilling team		X	
		Quarry platoon			X
		Real estate team			X
		Forward engineer support team-advanced*	X		X
		Forward engineer support team-main*	X		X
		Area clearance platoon		X	X
*Assets of the U.S. Army Corps of Engineers					
Legend: ARNG Army National Guard USAR U.S. Army Reserve					

Chapter 1

ORGANIC

1-15. The BEB commander is the senior engineer in the BCT and advises the maneuver commander on how best to employ combat, general, and geospatial engineering capabilities to conduct combined arms integration in support of decisive action, especially during early-entry operations. The BEB provides organic engineer planning and execution capabilities to the BCT. An armor BCT and an infantry BCT may have a brigade special troops battalion or a BEB during the transition to a BEB. A BEB will be formed in the Stryker BCT, where the engineer company is a separate company. For those units that have not transitioned to a BEB, refer to FM 3-34.22. The BEB has mission command of assigned and attached engineer companies, and the BEB is assigned a military intelligence company; a signal company; and a chemical, biological, radiological, and nuclear reconnaissance platoon (located in the headquarters and headquarters company). In the Stryker BCT, the battalion also provides mission command for an antitank company. The BEB is a comprehensive unit that provides maneuver support for bridging, breaching, route clearance, explosive hazard identification, and horizontal construction support. (See figure 1-2.)

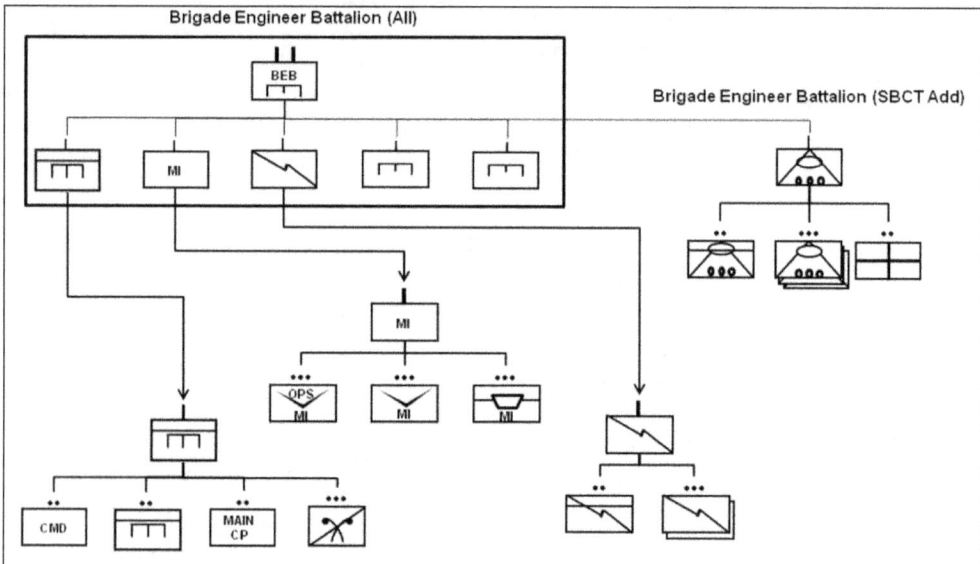

Notes.
1. The combat engineer companies shown are generic. Add the appropriate organizational icon to the basic function symbol for the brigade combat team affiliation.
2. An antitank company is added to the BEB for each SBCT organization.

Legend:	
BEB	brigade engineer battalion
cmd	command
CP	command post
MI	military intelligence
ops	operations
SBCT	Stryker brigade combat team

Figure 1-2. BEB

1-16. The BEB is responsible for administrative, logistical, training, and protection support of subordinate units. The BEB has a typical functional staff; however, the staff is predominantly engineers. The typical staff is as follows:
- **Human resources section.** The human resources section is responsible for the personnel administration of the many specialized military occupational skills of the battalion.
- **Military intelligence company.** The military intelligence officer in the intelligence section is primarily responsible for providing intelligence to the BEB and assisting the military intelligence company. The military intelligence company will receive administrative and sustainment support from the BEB.
- **Operations section.** The operations section includes combat, general, and geospatial engineers who will be at the center of technical planning and estimating. The operations section is responsible for training, operations, and plans for the battalion.
- **Chemical, biological, radiological, and nuclear platoon.** The chemical, biological, radiological, and nuclear platoon is responsible for providing technical advice to the BEB. The chemical, biological, radiological, and nuclear platoon will receive administrative and sustainment support from the BEB.
- **Sustainment section.** The sustainment section is responsible for coordinating the integration of supply, maintenance, transportation, and services for the battalion.
- **Signal company.** The signal company is primarily responsible for network management, knowledge management, and information assurance to the BEB. The signal company will receive administrative and sustainment support from the BEB.

1-17. The BCT commander will issue mission orders for these units. The command and support relationship dictates whether the BEB will logistically support or coordinate support with the BCT, brigade support battalion, or other unit higher headquarters. Unless the BCT directs otherwise, the BEB retains command and support relationships with organic and attached units, regardless of location on the battlefield. The companies may be further task-organized to maneuver task forces, the reconnaissance squadron, or a subordinate company or troop.

1-18. In some instances, the BEB may be directed to secure BCT command posts or execute security missions for areas that are not assigned to another unit in the BCT area of operations or the BEB may also be assigned responsibility for base camp defense, rear area defense, or terrain management. The BCT must weigh the risks associated with these missions because doing so would greatly diminish the BEB ability to operate as a functional headquarters, and it may reduce engineer support to the combined arms battalions and reconnaissance squadrons. To mitigate risk, the BEB staff may recommend additional engineer augmentation from EAB units. The BEB can defeat Level 1 threats and, with augmentation, organize response forces to defeat threats that are more organized.

1-19. Two engineer companies provide the BCT with the minimum capability to support offensive and defensive tasks (breach and cross obstacles, assist in the assault of fortified positions, emplace obstacles to protect friendly forces, construct or enhance survivability positions, conduct route reconnaissance and information collection, identify and clear improvised explosive devices) during decisive action. This maintains the BCT freedom of maneuver and inhibits the enemy ability to mass and maneuver. Each company is slightly different, but the primary focus is to support the combat engineering discipline with breaching, gap crossing, digging assets, and route clearance capabilities.

Engineer Company 1

1-20. Engineer company 1 (the number is notional to show the difference between the companies) is identical in the armor, infantry, and infantry (airborne) BCTs. This engineer company provides combat engineer support, and it consists of a company headquarters, two combat engineer platoons, and one engineer support platoon. The company provides mobility, countermobility, survivability, and limited construction support to the BCT. The combat engineer platoons provide the BCT with assets for breaching and obstacle emplacement. The engineer support platoon consists of a platoon headquarters; a horizontal squad that provides specialized engineer equipment to support limited general engineering tasks assigned to the company; and a breach squad that provides specialized equipment to support mobility, countermobility, and sustainment tasks assigned to the company. In a Stryker BCT, engineer company 1 has a company headquarters and two combat engineer platoons; but instead of an engineer support platoon, it has a bridge section. The breach squad of the Stryker BCT is limited to mine-clearing line charges and proofing equipment in the company. (See figure 1-3.)

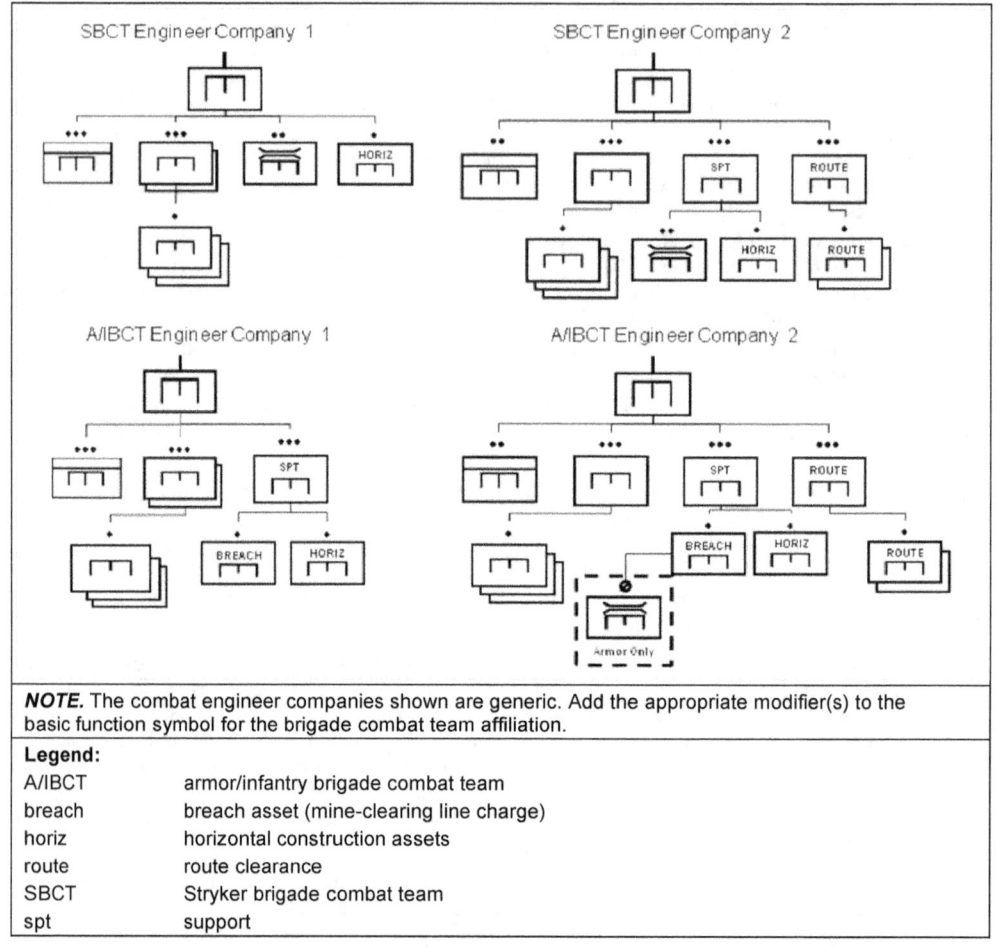

NOTE. The combat engineer companies shown are generic. Add the appropriate modifier(s) to the basic function symbol for the brigade combat team affiliation.

Legend:
A/IBCT   armor/infantry brigade combat team
breach   breach asset (mine-clearing line charge)
horiz   horizontal construction assets
route   route clearance
SBCT   Stryker brigade combat team
spt   support

Figure 1-3. Engineer companies 1 and 2

Engineer Company 2

1-21. Engineer company 2 (the number is notional to show the difference between the companies) is slightly different in the armor, infantry, infantry (airborne), and Stryker BCTs. Engineer company 2 is generally of the same composition as engineer company 1, but it has an additional route clearance platoon. This platoon provides the detection and neutralization of explosive hazards and reduces obstacles along routes that enable force projection and logistics. This route clearance platoon can sustain LOCs as members of the combined arms team or autonomously in a low-threat environment. The armor and infantry organization for this company is organized the same; however, the breach section contains different equipment and capabilities. The breach section consists of bridging, whereas, the infantry BCT and infantry (airborne) BCT breach section consists of mine-clearing line charges. The infantry BCT currently does not have bridging capability and will require augmentation from EAB engineers if the capability is required. (See figure 1-3.)

ECHELONS ABOVE BRIGADE

1-22. Engineer headquarters provide mission command for subordinate elements. Each has a staff that assists the commander in exercising mission command over subordinate engineer organizations and other selected nonengineer units to support multifunctional missions (combined arms breaching, combined arms gap crossing). The units in this category consist of the TEC, the engineer brigade, and the engineer battalion.

1-23. Baseline engineer units provide combat and general engineering capabilities that are primarily focused on the tactical to operational levels. Baseline engineer units are used to augment BCT engineers and to provide engineering capabilities to EAB engineer headquarters. When supporting a division or a corps, baseline engineer units are typically attached or under the operational control of the maneuver enhancement brigade (MEB) or the engineer brigade. When supporting echelons above corps, the baseline engineer units are attached or under the operational control of a functionally focused engineer brigade, the TEC, or the MEB.

1-24. Specialized engineer units are technically oriented (often low-density) units that provide specialized capabilities in construction support, infrastructure development, explosive hazards mitigation, geospatial support, well drilling, real estate management, military working dog units, prime power, technical rescue, diving, and firefighting. The specialized engineer units primarily support the operational to strategic levels, but they also provide selected support at the tactical level.

Engineer Headquarters

1-25. There are three echelons of engineer headquarters units—the TEC, the engineer brigade, and the engineer battalion. Multifunctional units may also provide mission command for engineer forces when engineer support is integral to the multifunctional mission. The engineer battalion is most often found in the engineer brigade, in the MEB, or in support of a BCT. The engineer brigade, one of the Army functional brigades, provides mission command for up to five engineer battalions at the division and corps levels. While not an engineer headquarters unit, the MEB is a significant headquarters for the employment of engineering capabilities. See FM 3-90.31 for additional information on the MEB.

1-26. The theater army normally receives one TEC. The TEC is designed to have mission command of assigned or attached engineer brigades and other engineer units within the supported geographic combatant command area of responsibility. When directed, it provides mission command for engineers from other Services and multinational forces and provides oversight of contracted construction engineers. The TEC focuses on operational level engineer support across the three engineer disciplines, and it typically serves as the senior engineer headquarters for a theater army, land component headquarters or, potentially, a joint task force (JTF). The TEC—

- Maintains primary responsibility for theater infrastructure development.
- Synchronizes engineer effort for the geographic combatant commander or JTF commander, providing peacetime training and support of military engagement for supported respective combatant commands.
- Deploys staff elements and organizations under CCDR authority.
- Establishes the primary purpose of fulfilling global operational requirements of an enduring and rotational nature to provide a wide range of technical engineering expertise and support on a per-mission basis.
- Consists of a command section and a deputy command section, and deploys its main command post and two deployable command posts.

1-27. The TEC commander receives policy guidance from the theater army based on the guidance of the geographic combatant command joint force engineer. In some theaters, a tailored engineer brigade may provide theater level engineer support. The engineer brigade provides expertise and capability that is similar to the TEC, but at a reduced level.

1-28. A division is not set by rules of allocation, and it is tailored to meet anticipated requirements based on mission analysis. The divisional engineer force may be organized under a multifunctional headquarters, such as the MEB, or it may be organized under a functional engineer brigade headquarters. In some situations, the division may require a combination of engineer forces that are organized under functional and multifunctional headquarters. The construct normally starts with a MEB and requires one or more engineer brigades when the number of engineer units, the technical nature of engineer missions, or the requirement to integrate engineering capabilities exceeds the span of control of a MEB.

1-29. Typically, an engineer brigade is allocated to a corps for most operations. The brigade can control up to five mission-tailored engineer battalions that are not organic to maneuver units. The battalions have capabilities from any of the three engineer disciplines. With augmentation, the engineer brigade may serve as a joint engineer headquarters and may be the senior engineer headquarters deployed in a joint operations area if full TEC deployment is not required. The engineer brigade headquarters—

- Provides mission command for assigned, attached, or operational control units of nonengineer units performing missions in support of a deliberate gap (river) crossing.
- Plans, supervises, and coordinates for combat engineer support, construction, facility rehabilitation, unit allocation, resource management, river crossing, barrier placement, countermine, and counterobstacle operations.
- Provides one deployable command post with the engineer staff expertise in technical planning, design, quality assurance and control, geospatial and terrain analysis, and the supervision of contract construction and labor.
- Provides support at a seaport of debarkation or an aerial port of debarkation (missions are terrain-focused) during early-entry operations or to support a movement corridor within a corps area of operations.

1-30. The engineer battalion provides mission command for one headquarters and headquarters company and one forward support company. The engineer battalion is assigned any variation of up to five engineer companies. When appropriately task-organized, it can provide mission command for combat and general engineering capabilities in support of a BCT, engineer brigade, or another unit. The engineer battalion can simultaneously support forces at all theater echelons. Due to habitual training relationships, some battalion headquarters are more capable in combat engineering than in general engineering or vice versa. Some battalion headquarters have additional capabilities (airborne, air assault, survey, design). The battalion may be focused on a single mission (route clearance, security, construction, cache interrogation, reduction). The engineer battalion may be organized to perform as a breach force command when the BCT is conducting a combined arms breach. During a gap-crossing (river-crossing) operation, the engineer battalion provides the option to be designated as the crossing-site command.

Baseline Engineer Units

1-31. Baseline engineer units include combat and general engineer units. The baseline engineer units are the primary building blocks for the organization of most engineer battalions. These units may augment the organic engineering capabilities of a BCT, or they may be task-organized under an engineer battalion headquarters to provide specific tailored capabilities to the EAB.

1-32. Baseline combat engineer units are focused on supporting combined arms operations at the tactical level. The baseline combat engineer units are designed to provide support to maneuver forces. Engineers have the capability to fight as engineers or, if required, as infantry. An engineer battalion headquarters is typically included to provide the necessary mission command, logistics, and staff supervision for attached and assigned combat engineer units when two or more are assigned to a BCT, MEB, or other organization. Combat engineer (sapper) units may construct tactical obstacles, defensive positions, and fixed and float bridges; repair command posts, tactical routes, culverts, and fords; and conduct general engineering tasks related to horizontal and vertical construction when augmented with the appropriate tools, equipment, and training. Combat engineer units also provide engineer support for gap-crossing operations, assist in assaulting fortified positions, and conduct breaching operations. Airborne and air assault-capable engineer units have the unique ability to employ air-droppable, rapid runway repair kits to support forcible, early-entry operations. The more specialized combat engineering capabilities of assault bridging, breaching, and route and area clearance are added to the organic engineering capabilities in BCTs (or to deployed baseline sapper companies) when required by the mission.

1-33. Baseline general engineer units include horizontal and vertical construction companies and engineer support companies. The baseline general engineer units construct, rehabilitate, repair, maintain, and modify landing strips, airfields, command posts, main supply routes, LOCs, supply installations, building structures, bridges, and other related aspects of the infrastructure. These units may also perform repairs and limited reconstruction of railroads or water and waste facilities. The basic capabilities of these units can be expanded by augmenting them with additional personnel, equipment, and training from specialized engineer units or other sources. Such augmentation can give them the capability to conduct bituminous mixing and paving, quarrying and crushing, pipeline construction, dive support, and major horizontal construction projects (highways, storage facilities, airfields).

Specialized Engineer Units

1-34. Specialized engineer units provide general and geospatial engineering capabilities at the operational and strategic levels, and the specialized engineer units augment those capabilities at the tactical level. Many capabilities are lower density than those of the baseline engineer units. These smaller, more specialized units are designed to support technical aspects within larger, engineer-related missions or to augment selected headquarters elements.

1-35. The engineer diving detachment performs scuba and surface diving operations to a depth of 190 feet in a maritime environment in support of combat, general, and geospatial engineering. Divers provide reconnaissance, river-crossing, hydrographic survey, demolition, port construction and rehabilitation, harbor clearance, ship husbandry, salvage, joint logistics-over-the-shore, and hyperbaric life support operations. (See ATTP 3-34.84 and TM 3-34.83 for additional information.)

1-36. Explosive hazard support provides the commander with specialized capabilities and integrates the tasks conducted to counter the explosive hazards threat. These capabilities include the linkage to Army explosive ordnance disposal (EOD) capabilities found in the ordnance branch. The engineer squad (canine) includes specialized search dog teams and mine dog teams. These teams assist in locating firearms, ammunition, and explosives in rural and urban environments. The teams may be used to augment a variety of route and area clearance capabilities found in the clearance company.

1-37. Construction support provides mission command for the management, procurement, and oversight of contracted support. It also enhances performance for asphalt, concrete, and haul operations. Each of these capabilities has a role in infrastructure support.

1-38. Infrastructure support includes the following capabilities:
- Engineer prime power units that generate electrical power and provide advice and technical assistance on all aspects of electrical power and distribution systems. Prime power units have limited electrical engineering capability (design and analysis); provide electrical surveys; and operate, maintain, and perform minor repairs to other electrical power production equipment, to include HN fixed plants.
- Engineer facility detachments that support theater opening and closing, base camp development, construction management, contract technical oversight, base camp operations (to include waste management functions), and master planning.
- Firefighting teams that provide base and base camp fire protection and search and rescue.
- Engineer technical rescue teams that provide first responder support for facilities, aviation, and tactical vehicle extraction.

1-39. Two specialized engineer units provide geospatial engineering capabilities—the topographic engineer company and the geospatial planning cell. Currently, the topographic engineer company provides geospatial support to deployed units that require augmentation. The topographic engineer company provides modules tailored to support the geographic combatant command; JTF headquarters; theater army, corps, and division headquarters; sustainment brigades; other joint or multinational division- and brigade-size units; and the Federal Emergency Management Agency regions. Geospatial engineering capabilities include analysis, collection, generation, management, finishing, and printing. Geospatial planning cells generate, manage, and disseminate geospatial data, information, and products in support of ASCC headquarters and geographic combatant commands. The intended goal of these organizations, in cooperation with organic geospatial teams, is to apply the relevant geospatial information available, to explain the military significance of the terrain and other spatial and temporal aspects of the operational environment to the commander, and to facilitate informed decisionmaking. Furthermore, these geospatial organizations conduct the exploitation of geospatial information and services, producing spatially accurate products for mission command, intelligence, measurements, mapping, visualization, and modeling.

1-40. Although the Army has only one dedicated engineer reconnaissance unit (in the combat engineer company of the armor BCT), commanders routinely form mission-tailored engineer reconnaissance teams to collect engineer-specific tactical and technical information. These teams are a critical source of information for engineers and combined arms commanders, playing an important role in the intelligence preparation of the battlefield. FM 3-34.170 provides a detailed discussion on the range of engineer reconnaissance capabilities.

Other Capabilities

1-41. The technical rescue company is a unique capability that exceeds the knowledge, skills, and equipment associated with firefighters and other emergency responders to resolve unique or complex rescue situations safely. The technical rescue company receives specialized training in structural collapse, trench, and tunnel rescue to provide immediate lifesaving treatment, stabilization, and extraction.

ENGINEER FORCE TAILORING

1-42. The organization of forces within the Army is dynamic. Actual requirements for forces are seldom identical to planning figures. Therefore, the theater army commander recommends the appropriate mix of forces and the deployment sequence for forces to meet the geographic combatant command requirements. This is called *force tailoring* (selecting forces based on a mission and recommended deployment sequence) and may include elements from the operational Army and the generating force.

1-43. Tailoring the engineer force requires a different mind-set—one that thinks in terms completely divested from how the force is organized in a garrison. It requires a leader to think beyond garrison structures and embrace combinations of modular engineering capabilities and scalable mission command to provide each echelon of the force with the right support. While the Engineer Regiment is organized and equipped to support unified land operations, engineers can expect serious challenges in the operational environment when trying to execute the broad range of potential tasks. Careful prioritization must occur for the limited engineer resources typical in the operational environment. To accomplish the identified tasks in the desired timeframes, commanders must consider augmentation requirements and recognize which mission requirements can be supported through reachback and geospatial products rather than enlarging the engineer footprint in the area of operations. Engineer units are more narrowly designed to accomplish specific types of tasks. Therefore, when tailoring the engineer force, it is imperative that a broad range of capabilities is allocated from the engineer force pool.

1-44. Engineer force packages must contain the right mix of capabilities to assure timely and relevant engineer support to the joint force command. This mix will often need to change drastically during transitions, and the joint force engineer must anticipate and plan for these changes. For example, combat engineers often make up the majority of engineer forces in-theater during sustained combat operations. However, combat engineers must be reinforced during transition to operations that are dominated by stability tasks, because they typically do not have the right capabilities to accomplish the required general engineering tasks. Also, since EOD support requirements during transitions are often significantly higher than during combat operations, more EOD capabilities will be required.

1-45. Tactical and operational commanders task-organize groups of units for specific missions. They reorganize for subsequent missions when necessary. This process of allocating available assets to subordinate commanders and establishing command and support relationships is called *task organizing*. Considerations for task-organizing engineer units are discussed in ATTP 3-34.23.

1-46. The execution of, and adherence to, the Army force generation model is problematic for engineer units because there is—
- A limited amount of engineer force structure compared to the Army BCT structure and potential BCT mission requirements.
- A minimal number of engineers organic to BCT organizations.
- A high percentage of engineer forces resident in the Reserve Components.

1-47. The implications of Army force generation on the engineer force are similar to those on other support branches within the Army where a majority of forces are not organic to a BCT structure. Activating an engineer unit early in the Army force generation process will have secondary and tertiary effects for operational, sustainment, and personnel planners. It reduces the availability of units later in the cycle. A surge of engineer units can be accomplished for short periods, but not indefinitely without looking at increasing engineer units in the inventory or using HN or contract engineers. Engineers are typically employed in modules, units, or companies, but are deployed in a battalion level organization.

UNITED STATES ARMY CORPS OF ENGINEERS

1-48. USACE is the Army direct reporting unit with assigned responsibility to execute Army and DOD military construction, real estate acquisition, and the development of the nation infrastructure through the civil works program. Other services include wetlands and waterway management and disaster relief support operations. (USACE has primary responsibility to execute the Emergency Support Function 3–Public Works and Engineering Course for DOD.) Most USACE assets are part of the generating force, but selected elements support the operational Army, to include various field force engineering (FFE) teams and the 249th Engineer Battalion (Prime Power). With its subordinate divisions, districts, laboratories, and centers, USACE provides a broad range of engineer support to military departments, federal agencies, state governments, and local authorities in a cost-reimbursable manner. USACE districts provide the design, contract support, construction, and operation of hydroelectric power generation plants and river navigation systems while reducing the overall environmental impact. USACE also provides technical assistance and contract support to joint forces that are deployed worldwide. USACE operates the U.S. Army Engineer Research and Development Center, a comprehensive network of laboratories and 43 centers of expertise. (See ATTP 3-34.23 for additional information.) The center includes the following facilities:
- Geotechnical and Structures Laboratory.
- Coastal and Hydraulics Laboratory.
- Environmental Laboratory.
- Information Technology Laboratory.
- Topographic Engineering Center.
- Cold Regions Research and Engineering Laboratory.
- Construction Engineering Research Laboratory.
- Army Geospatial Center.

MISSIONS

1-49. USACE capabilities include access to the expertise of the U.S. Army Engineer Research and Development Center laboratories and centers and the resources within the divisions, districts, and other sources. USACE has aligned its divisions with, and assigned liaison officers to, combatant commands and selected ASCCs to enable access to USACE resources to support engagement strategies and wartime operations. The USACE mission supports unified land operations with the following major functions:
- Water resource functions support the balance between water resource development and environmental impact.
- Infrastructure functions acquire, build, and sustain critical facilities for military installations, theater support facilities, and public works.
- Environmental functions restore, manage, and enhance local and regional ecosystems.

Chapter 1

- Research and development functions work toward the innovation, introduction, and improvement of products and processes in support of the warfighter; installations; and energy, environmental, and water resources.
- Civil disaster response functions respond to and support recovery from local, national, and global disasters.
- Military contingencies provide engineering and contingency support for unified land operations.

1-50. USACE provides technical and contract engineer support, integrating its organic capabilities with those of other Services and other sources of engineer-related reachback support. USACE may also have assets directly supporting separate commands, the TEC, or senior engineer headquarters in-theater. Whether providing engineer planning and design or contract construction support in the area of operations or outside the contingency area, USACE can obtain the necessary data, research, and specialized expertise that is not present in-theater or through reachback capabilities, using teleengineering when necessary. Teleengineering is the communications architecture that facilitates reachback when the existing communications infrastructure will not support it. Teleengineering is under the proponency of the USACE and is inherent in FFE.

FIELD FORCE ENGINEERING

1-51. USACE aligns its divisions with specific combatant commands. A USACE division integrates USACE capabilities to meet combatant command requirements and provide mission command for USACE activities in the area of operations. USACE supports all combatant commands. (See table 1-2 for USACE division alignment.)

Table 1-2. USACE division alignment

USACE Division	Supported Combatant Command
North Atlantic Division	U.S. Africa Command
	U.S. European Command
Transatlantic Division	U.S. Central Command
Northwestern Division	U.S. North Command
Pacific Ocean Division	U.S. Pacific Command
South Atlantic Division	U.S. South Command
Legend: USACE U.S. Army Corps of Engineers	

1-52. USACE is the primary proponent of FFE and related generating-force support that enables engineer support to the operational Army. ***Field force engineering* is the application of the Engineer Regiment capabilities from the three engineer disciplines (primarily general engineering) to support operations through reachback and forward presence.** FFE is provided by technically specialized personnel and assets (deployed or participating through the USACE reachback process or through operational force engineer Soldiers linked to reachback capabilities). The engineer commander maintains flexibility and determines the mix of capabilities (troop, USACE civilian, and contractor) based on the tactical situation, time-phased requirements, capabilities required, available funding, and force caps. The USACE division commander task-organizes division capabilities to meet the varying time-phased requirements. These capabilities rely heavily on reachback through communication systems (teleengineering). The FFE concept is applicable in joint and multinational operations to provide technical engineer solutions that can be implemented expeditiously and with a small footprint forward. The U.S. Air Force and U.S. Navy have similar capabilities—the Air Force uses its Geo-Reach Program, while the Navy has the capability to conduct engineer reconnaissance with reachback to the Naval Facilities Engineering Command.

1-53. USACE objectives for FFE are to—
- Leverage its generating-force capabilities (engineering expertise, contract construction, real estate acquisition and disposal, environmental engineering, and waterways management) in operations.
- Maximize the use of reachback to provide technical assistance and enable operational force engineers in support to the CCDR.

1-54. One of the ways USACE accomplishes these objectives is by training, equipping, and maintaining specialized, deployable FFE teams. These deployable USACE organizations provide technical assistance, enable operational force engineers, and access additional technical support through reachback. Another way that USACE supports the operational force is through nondeployable teams that provide dedicated engineer assistance in response to requests for information from deployed teams or engineer Soldiers in the operational area. Focus areas for these teams include infrastructure assessment and base camp development. Liaison officers are provided to the geographic combatant commands and select ASCCs (plans and operations) on a full-time basis. These liaison officers communicate and integrate USACE capacity into combatant commands and ASCCs, and they provide USACE headquarters and major subordinate commands with situational awareness with a focus toward impending or ongoing USACE operations in support of the combatant commands or ASCCs.

1-55. The FFE teams and the USACE Reachback Operations Center are the primary contacts within USACE that are organized, trained, and equipped to provide technical solutions to engineer and construction-related challenges. FFE teams deliver technical engineer support to supported units through engineer staff. FFE teams provide embedded engineer planning and technical engineer support to unified land operations or offer dedicated reachback support to deployed teams and engineer Soldiers in need of technical support. FFE teams typically develop solutions by employing available resources, but the teams have the option to employ reachback to the entire array of expertise within the USACE laboratories or centers of expertise for more complex engineering issues. USACE has expertise that may support the strategic, operational, or tactical level in engineer planning and operations. USACE can leverage reachback to technical subject matter experts in districts, divisions, laboratories, and centers of expertise; other Services; and private industry as part of the USACE role in generating force. FFE is a means to access specialized engineering capabilities that can augment joint forces command planning staffs.

Forward-Deployed Units

1-56. Teams can rapidly deploy to meet requirements for engineering assessments and analyses in support of the full array of engineer missions. Teams include forward engineer support teams (FESTs), contingency real estate support teams, and environmental support teams.

Forward Engineer Support Team-Advance

1-57. A forward engineer support team-advance (FEST-A) is a deployable team that provides infrastructure assessment; engineer planning and design; and environmental, geospatial, and other technical engineer support (from theater army to brigade echelon) and augments the staff at those echelons. This team could support any echelon configured as a joint force headquarters for limited contingency operations or may be task-organized at corps, division, and brigade echelons when configured as an intermediate or tactical headquarters. The FEST-A operates as augmentation to the supported force engineer staff or to the supporting engineer headquarters. It is designed to receive a contingency real estate support team and an environmental support team when those capabilities are required. The FEST-A conducts a variety of core essential tasks in support of stability operations, incident management or DSCA, and technical engineering missions. The FEST-A consists of uniformed military personnel and DA civilians that require sustainment and security support from the gaining or supported unit.

Forward Engineer Support Team-Main

1-58. A forward engineer support team–main (FEST-M) is a deployable team that provides construction management and real estate, environmental, geospatial, and other engineer support (typically to the theater army) and can provide mission command for deployed FFE teams. FEST-Ms are not required for initial entry into a theater of operations; therefore, USACE will organize, train, and equip FEST-M units should the requirement develop during a contingency. This team would typically support a JTF or the land component of a JTF. The FEST-M is task-organized to that headquarters or to a supporting engineer headquarters. The FEST-M operates as augmentation to the joint force engineer staff or the engineer headquarters element or may operate as a discrete headquarters element. It is designed to provide mission command for additional FFE elements when task-organized with those organizations. In some cases, the FEST-M may provide the foundation upon which a contingency engineer district is established in-theater. The FEST-M element also conducts a variety of core essential tasks in support of stability operations, incident management or DSCA, and technical engineering missions. It requires sustainment and security support from the gaining or supported unit.

Contingency Real Estate Support Team

1-59. A contingency real estate support team is a deployable team that can acquire, manage, and oversee the disposal of real estate on behalf of the U.S. government pursuant to delegated authority under Section 2675.3, Title 10, U.S. Code (10 USC 2675), and specific delegation from the Office of the Deputy Assistant Secretary of the Army Installations, Housing, and Partnerships. This team can support any echelon, but it is typically tailored to support an Army component headquarters configuration with support missions requiring real estate management. The contingency real estate support team operates as augmentation to the supported force engineer staff or supporting engineer headquarters. It may also be task-organized to a tailored FEST. The contingency real estate support team conducts real estate management tasks, and it should be deployed early in a contingency to facilitate the acquisition of real estate in support of the development of facilities for U.S. forces. It requires sustainment and security support from the gaining or supported unit.

Environmental Support Team

1-60. An environmental support team is a deployable team that conducts environmental assessments, baseline studies, and other surveys and studies. This team can support any echelon, but it is typically tailored in support of an Army component headquarters configuration with support missions requiring base camp development. The environmental support team operates as augmentation to the supported force engineer staff or to the supporting engineer headquarters. It may also be task-organized to a tailored FEST. The environmental support team conducts environmental management tasks in support of base camps and other technical engineering missions. The team should be deployed as an initial capability to perform assessments, identify environmental hazards, and remain as one of the last to provide remediation actions and support for base or base camp closure. The environmental support team requires sustainment and security support from the gaining or supported unit.

Reachback Support

1-61. The USACE Reachback Operations Center is the reachback capability to the USACE for technical engineering requests for information. The USACE Reachback Operations Center mission is to provide rapid, relevant, and reliable solutions to Soldiers and civilians in support of the armed forces and the nation. The USACE Reachback Operations Center supports the warfighter and the nation by providing cost effective, superior customer service and by achieving customer (requestor) satisfaction in every area of reachback support. This reachback engineering capability allows U.S. personnel who are deployed worldwide to talk directly with experts when a problem in the field needs quick resolution. Deployed personnel can be linked to subject matter experts within the U.S. government, DOD, USACE, private industry, and academia to obtain a detailed analysis of complex problems that would be difficult to achieve with the limited expertise or computational capabilities available in the field. While the USACE Reachback Operations Center is capable of responding to a variety of complex technical problems, the team is also trained to exploit the entire array of expertise within USACE laboratories, centers of expertise, base camp development teams, USACE divisions and districts, other DOD or U.S. government agencies, or other organizations for more complex engineering issues.

1-62. A base camp development team is a nondeployable team within a selected USACE district that can quickly provide base camp development engineering, master planning, and facilities design in support of FFE and other reachback requests for information. Base camp development teams are trained and organized within USACE divisions, and they maintain a rotational readiness cycle. While these teams are capable of responding to a variety of complex technical problems, they are also trained to exploit the entire array of expertise within USACE laboratories or centers of expertise for more complex engineering issues. Focus areas for the base camp development teams are engineer-related planning and development issues involved in locating, designing, constructing, and eventually closing or transferring base camps. Base camp operations and maintenance activities are not within the scope of FFE support, but may rely on FFE applications to address specific technical engineering requirements when necessary. The base camp development team resources and expertise are available to support FFE teams and operational forces through the USACE Reachback Operations Center Web site.

1-63. A USACE contingency engineer district can be employed to support combatant command requirements. Should contract construction exceed the capabilities of the major subordinate commander, USACE may establish an additional contingency support organization or simply augment the existing staff. These capabilities would align themselves with the theater level engineer staff element.

1-64. The USACE contingency engineer district mission is to provide responsive technical engineer support to U.S. forces, other U.S. government agencies, and headquarters and staff augmentation. Technical engineer support includes—
- Engineer reconnaissance.
- Estimates.
- Design and plan projects.
- Execution of contract construction.
- Project quality control.
- Real estate acquisition and disposal.
- Environmental assessments and operations.
- Operation and maintenance until projects are turned over to designated agencies.
- Technical engineering advice to the supported command and agencies in support of the operational campaign plan.

1-65. USACE contingency engineer district business management functions include—
- Project contracting.
- Resource management operations.
- Legal support for contract construction.
- Safety and occupational health.
- Logistics management operations.

1-66. The headquarters and staff augmentation cell is a USACE cell that synchronizes USACE construction effects and customer requirements at the theater level, provides oversight of USACE construction programs, and provides USACE with major subordinate command level of oversight of forward to deployed contingency engineer districts and other USACE assets. This augmentation cell provides the theater-wide synchronization of construction effects and program oversight. The augmentation cell roles and responsibilities include—
- Communicating construction effects to theater level decisionmakers and jointly developing an infrastructure plan that nests with the civil-military strategy as part of the theater campaign plan.
- Augmenting theater engineer staff with technical engineering expertise not resident within the staff.
- Engaging with HN government and the U.S. government interagency to synchronize construction effects.
- Pulling information from contingency engineer districts and synthesizing it into construction effects for reporting purposes and providing programmatic oversight of construction programs.
- Advising the supported command on USACE core competencies and supporting construction decisions for work acceptance.
- Participating in the development of infrastructure construction requirements and providing strategic command guidance to deployed contingency engineer districts according to theater and USACE strategic command objectives.

- Providing oversight of the USACE major subordinate commander with the responsibility to support the geographic combatant command and being the operational conduit between deployed elements and the USACE major subordinate commander.
- Pushing guidance to the deployed contingency engineer districts from the supported theater command.
- Prioritizing mission execution and ensuring the implementation of USACE business processes consistently throughout the theater.
- Acting as the USACE major subordinate commander (forward) for the USACE major subordinate commander with global combatant commands responsibility and managing USACE affordability in the theater of operations.

Note. See Engineer Pamphlet 500-1-2 for additional information on FFE.

UNIFIED ACTION PARTNERS

1-67. Military engineers may need to coordinate activities with other nation forces, U.S. government agencies, nongovernmental organizations, United Nations agencies, and HN agencies according to the operational mandate or military objective. In all cases, the authority must exist for direct coordination. Military engineers must establish interagency relationships through negotiation. The specific agency will vary, depending on who has federal or state proponency for the situation (for example, disaster relief versus a firefighting mission). Agreements should be written as memorandums of understanding or terms of reference to ensure understanding and avoid confusion. Most agreements are made at the combatant command or JTF level and normally place serious legal restrictions on using military personnel and equipment. These agencies and organizations may have unique engineering capabilities that could be used as part of the overall operational effort. However, these agencies and organizations often request extensive engineer support of activities and programs. It is critical that an effective engineer liaison is established with the force headquarters civil-military operations center to coordinate and execute any engineer support exchanged with these agencies.

JOINT CONSIDERATIONS

1-68. Army engineers frequently operate in a joint environment and must understand joint command and support authorities and relationships (described in JP 1), which are similar, but not identical to Army command and support relationships. They must understand how these are applied in joint engineer operations as described in JP 3-34. Particularly pertinent to engineer operations are—
- The directive authority for logistics that CCDRs have and authority to delegate directive authority for common support capabilities, which include engineer support.
- The authority to employ mines, which originates with the President. (See JP 3-15 for additional information.)

1-69. Army special operations forces provide an array of formations that are capable of rapidly reversing the conditions of human suffering by decisively resolving conflicts. Engineers support Army special operations forces through a number of unique capabilities and tasks that include geospatial information and services, infrastructure development, facility construction and maintenance, training an indigenous population on how to construct protective obstacles, supply mobile electric power, and facility hardening. Special operations support can be performed at the company, platoon, squad, or Soldier level. Support to special operations tends to require smaller elements with multifunctional capability. Contracting, logistics, and engineering operations work hand in hand throughout the special operations area of responsibility.

1-70. Cyber electromagnetic activities are continuous and unimpeded by geography. This domain leverages the electromagnetic spectrum through wireless systems. Wireless systems are enablers to modern telecommunications, computer networks, and weapon systems. Engineers are enablers and users of cyber electromagnetic activities. Engineers support these activities through tasks that include—
- Hardening facilities.
- Constructing protective obstacles.
- Providing uninterrupted medium-voltage electrical power.
- Providing clean and secure power/energy supply and grid systems to mitigate and minimize cyber electromagnetic disruptions that may have been caused by adversary systems. Disruptions of cyber electromagnetic activities will affect geospatial data, global positioning devices, and data from sensors and mines.

1-71. Engineers rely on space-based capabilities and systems (global positioning systems, communication and weather satellites, intelligence collection platforms) to be successful in combat, general, and geospatial engineering. The planning and coordination of space support with national, Service, joint, and theater resources takes place at the corps and division levels to provide expertise, advice, and planning that may directly affect and impact engineer tasks to plan, communicate, maneuver, and maintain situational awareness; conduct reconnaissance; and protect and sustain U.S. forces. Space-enabled capabilities are widely used to maintain situational awareness. Space-based systems are critical during engineer operations because they—
- Provide rapid communication that enables a commander to gain and maintain the initiative by developing the situation faster than the enemy by visualizing the battlefield, sharing a common operational picture, retaining the ability to recognize and protect U.S. and friendly forces, synchronizing force actions with adjacent and supporting units, and maintaining contact and coordination.
- Provide communication links between forces and commanders within the theater.
- Provide updates of the solar environment and its impact to terrestrial and space-based segments of friendly communication systems.
- Monitor terrestrial areas of interest through information collection assets to help reveal the enemy location and disposition and route, area, zone, and force reconnaissance.
- Provide global positioning system status and the accuracy of positioning, navigation, and timing for planning and conducting geospatial engineering.
- Provide meteorological, oceanographic, and space environmental information that is processed, analyzed, and leveraged to produce timely and accurate weather effects and impacts on operations.

1-72. The U.S. Army Strategic Forces Command Commercial Imagery Team performs a complementary geospatial information and support mission. It provides commercial imagery data and products to customers in a multinational environment. The team consists of space experts, a satellite communications control technician, terrain data experts, topographic analysts, and an information systems specialist. They produce many different imagery products (image maps, change detection, terrain categorization, multispectral analysis), depending on the needs of the customer. The U.S. Army Strategic Forces Command Commercial Imagery Team coordinates with geospatial support teams, collection managers, and commercial imagery vendors to provide multinational forces with a releasable, unclassified commercial satellite geospatial product.

MULTINATIONAL CONSIDERATIONS

1-73. NATO and the American, British, Canadian, Australian, and New Zealand Armies Program engineering capabilities are well known, and data about them is readily available. (See FM 3-16 for additional information on multinational operations.) Standardization agreements between national armies facilitate engineer interoperability and cooperation. The capabilities of engineers from other nations are normally available through intelligence channels or formal links with the nations concerned. Several nations have engineers that are experts in specific combat engineering tasks (mine detection and removal). Other national engineers are focused on specific missions (disaster relief). Engineers must have an appreciation for the engineering capabilities and limitations of other nations. AJP 3.12(A) provides a necessary starting point for working with allied engineers.

1-74. Depending on the multinational force arrangement in-theater, Army engineers may control or work closely with engineers from other nations. Command and support relationships for multinational engineer forces are established to foster the unity of effort. Providing adequate U.S. engineer liaison officer support (linguist support, communications equipment, transportation) is critical to this process.

1-75. During force projection operations, the initial engineering capabilities in-theater can employ a mix of HN, contracted, and multinational capabilities. As Army engineers deploy into a theater, they may be joined by multinational and joint engineers. When coordinating multinational engineer plans and operations, the theater army engineer staff should consider the joint considerations that are addressed in JP 3-34 and the following:

- Requesting the latest intelligence information concerning the HN or multinational engineer structures and logistics requirements.
- Requesting the latest engineer intelligence data from the HN or deploying multinational engineer elements to help identify force projection theater army engineer requirements and enemy engineering capabilities. Requirements include explosive hazard and obstacle data; soils data; construction materials availability; HN construction support; and terrain, infrastructure, and climate data.
- Establishing multinational engineer staff links between the theater army, HN, and multinational engineer force staff sections through the JTF or combatant commands engineer staff and headquarters.
- Providing NATO multinational command and control with the NATO operation order (OPORD) format and the NATO decisionmaking process.
- Providing necessary Army engineer liaison officer support.
- Developing the multinational task organization relationships that enhance HN and multinational engineering capabilities following the deployment of Army engineers.
- Assessing the need for HN and multinational engineer support following the arrival of Army engineer units in-theater.
- Determining if multinational engineer units need augmentation from Army engineer units.
- Developing procedures for Army engineer units to support multinational engineers with additional Class IV construction materials and engineer equipment.

INTERAGENCY CONSIDERATIONS

1-76. Interagency cooperative agreements expand the scope and capabilities of any given response because of the wide variety of expertise and funding resources that are potentially available. Not only do interagency operations increase the resources engaged in an operation, but they also increase and complicate the coordination necessary to conduct operations. Engineer support to operations may be significantly impacted by the participation of interagency organizations. Engineer support may be a key enabler to these operations. During the conduct of stability operations, interagency organizations will employ contract or other construction capabilities concurrently with ongoing military engineer support. Coordination can identify and avoid conflicting issues and unify the effect of these efforts. The following interagency organizations could be involved:

- Federal Emergency Management Agency.
- Environmental Protection Agency.
- Drug Enforcement Administration.
- National Oceanic and Atmospheric Administration.
- U.S. Geological Survey.
- Public Health Service.
- Civil Air Patrol.
- Department of Agriculture.
- Department of State.

- U.S. Agency for International Development.
- Office of Foreign Disaster Assistance.
- Department of the Interior, Fish, and Wildlife Agency.
- General Accounting Office.

HOST NATION CONSIDERATIONS

1-77. In a forward-deployed theater, the theater army identifies wartime facility and construction requirements for the Army as part of the deliberate war planning effort. Construction requirements are identified using the planning module in the theater construction management system. Subsequent analyses further refine construction requirements and provide a basis for—

- Force structuring.
- Procurement.
- Lease provisions and HN agreements.

1-78. The product of these analyses is the engineer support plan. The goal is to reach HN support agreements in peacetime to provide maximum facilities within the theater. Advanced planning and the commitment of resources by HNs reduce the early lift requirements needed to support reception, staging, onward movement, and integration. Written agreements with HNs regarding support items foster an understanding of the assistance levels and increase the likelihood of execution. Engineer support from the HN usually involves providing—

- Land.
- Facilities.
- Construction support.
- Manpower.
- Equipment.
- Materials.
- Services.
- Waste disposal.

1-79. Wartime HN support agreements in forward-presence theaters (Europe and Korea) have been negotiated to provide HN construction support (facility modifications, LOC maintenance and repair, utility services). During contingency operations, HN support agreements tend to be less formal; however, these agreements are no less critical to mission success in the event of an operation. Such HN support is used when possible to free U.S. engineer units for critical missions where HN support alternatives are not viable. Support agreements are negotiated in peacetime on an asset basis. Assets may be facilities, contracts, or equipment. Again, this support is particularly critical during the initial stages of a contingency when reception, staging, onward movement, and integration requirements are high and engineer assets are limited.

1-80. Pre-positioning engineer equipment within the region reduces the response time into a particular theater by allowing engineer forces to deploy by air and fall in on war stocks within the region. These pre-positioning locations are a critical element of the U.S. force projection national strategy and represent a significant contribution of HN support. Beyond direct HN support, multinational elements directly or indirectly involved in the crisis may provide other support. Other nations sympathetic to the cause may be limited in direct participation because of constitutional restrictions or political sensitivities. However, these nations may provide engineer equipment, supplies, or funding, much like the Japanese provided during the Gulf War.

1-81. During a conflict, the HN may provide local contractors to repair or construct facilities. Construction materials (cement, asphalt, aggregate, timber, steel) and contract labor may also be available. HN assets may also be available for providing local security and for transporting construction materials and equipment. Third world country nationals may be available by request through the HN or direct contact with nationals to support engineer activities. Engineer reconnaissance and assessment teams engaged in planning during peacetime or dispatched early in contingency operations are the key to identifying and accessing available HN assets.

Nongovernmental Organizations

1-82. Relationships with international and domestic nongovernmental organizations must be established through negotiation. Most agreements are made at the strategic level; however, the operational and tactical commanders may have some latitude delegated to them. Agreements normally have serious legal restrictions on using military personnel and equipment. Some of these agencies may have unique and significant engineering capabilities that could be used as a part of the overall operational concept. These capabilities may be a useful source of Class IV supplies, not only for agency projects, but also as a negotiated barter for services rendered in support of its mission. More often than not, however, these agencies and organizations may request extensive engineer support for activities and programs. As these organizations play an important part in the CCDR achievement of strategic objectives, the demands must be coordinated. Therefore, it is critical that an effective engineer liaison be established and maintained with the force headquarters civil-military operations center.

1-83. The United Nations may designate a regional organization, which has a greater vested interest and appreciation for the forces at work in a given region, as its operational agent to exercise control. These organizations have different operational concepts and organizational procedures. U.S. forces are familiar with some of these concepts and procedures (such as NATO), but are not familiar with others.

Chapter 2

Engineer Support to Unified Land Operations

Army engineer support to operations encompasses a wide range of tasks that require many capabilities. Commanders use engineers throughout unified land operations across the range of military operations. They use them primarily to assure mobility, enhance protection, enable force projection and logistics, and build partner capacity and develop infrastructure. This chapter describes engineer tasks, the lines of engineer support, and engineer support to the warfighting functions.

ENGINEER TASKS

2-1. Engineer tasks provide the freedom of action as the objective. Engineer tasks that affect terrain deal with obstacles (including gaps), bridges, roads, trails, airfields, fighting positions, protective positions, deception, and a wide variety of other structures and facilities (base camps, aerial ports, seaports, utilities, buildings). Engineers affect these by clearing, reducing, emplacing, building, repairing, maintaining, camouflaging, protecting, conserving, or modifying them in some way through tasks (obstacle reduction, route clearance, technical rescue, infrastructure and environmental assessments, geospatial engineering).

2-2. Regardless of category, engineer tasks have different purposes in different situations. For example, a task to clear explosive hazards from a road that is designated as a direction of attack may have the purpose of assured mobility. Two days later, that same road may be designated as a main supply route, and a task to clear explosive hazards from the road may have the purpose of protecting critical assets or enabling logistics. The same task is involved, but with different purposes. In addition to the different purposes that an engineer task can have at different times, engineer support often involves simultaneous tasks with different purposes that support different warfighting functions. This chapter explains how engineer tasks are grouped by purpose into the lines of engineer support, how they are grouped into the types of operations, and how they contribute to the warfighting functions.

LINES OF ENGINEER SUPPORT

2-3. Fundamental to engineer support to operations is the ability to anticipate and analyze the problem and understand the operational environment. Based on this understanding and the analysis of the problem, engineer planners select and apply the right engineer discipline and unit type to perform required individual and collective tasks. They must think in combinations of disciplines, which integrate and synchronize tasks in concert with the warfighting functions to generate combat power. Finally, they establish the necessary command and support relationship for these combinations. The lines of engineer support are the framework that underpins how engineers think in combinations, and these lines provide the connection between capabilities and tasks.

2-4. Commanders use lines of engineer support to synchronize engineer tasks with the rest of the combined arms force and to integrate them into the overall operation throughout the operations process. Lines of engineer support are categories of engineer tasks and capabilities that are grouped by purpose for specific operations. Lines of engineer support assist commanders and staffs with the capabilities of the three engineer disciplines throughout the Engineer Regiment and align activities according to purpose. The engineer disciplines are capabilities, based on knowledge and skills, that are organized in units. These units are organized based on the disciplines that are executed through individual and collective tasks. The combination of these tasks for a specific purpose, in the context of decisive action, achieves the lines of engineer support.

Chapter 2

2-5. Regardless of where a task falls within the Army universal task list, task alignment with a line of engineer support is determined by the purpose of the task in a given situation. Engineer support is primarily focused on achieving the four lines of engineer support.

2-6. The three engineer disciplines encompass tasks along the lines of engineer support. The combat engineering discipline, due to its support to maneuver forces in close combat, is primarily focused on tasks that assure mobility and enhance protection. The general engineering and geospatial engineering disciplines performs tasks along all four lines of engineer support.

ASSURE MOBILITY

2-7. The assure mobility line of engineer support orchestrates the combat, general, and geospatial engineering capabilities in combination to allow a commander to gain and maintain a position of advantage against an enemy (mobility operations) and deny the enemy the freedom of action to attain a position of advantage (countermobility operations). These tasks primarily focus on support to the movement and maneuver warfighting function, to include support to special operations forces. Although normally associated with organic combat engineers, general engineers may be task-organized to support this line of engineer support.

2-8. This line of engineer support does not include engineer tasks intended to support the administrative movements of personnel and materiel. Such tasks are normally intended to enable logistics. The assure mobility line of engineer support is achieved through the assured mobility framework described in ATTP 3-90.4.

Support to Mobility Operations

2-9. Engineer support to mobility operations includes the following primary tasks:
- Conducting combined arms breaching.
- Conducting area and route clearance.
- Conducting gap crossing.
- Constructing and maintaining combat roads and trails.
- Constructing and maintaining forward airfields and landing zones.
- Conducting traffic management and enforcement.

2-10. *Mobility operations* are those combined arms activities that mitigate the effects of natural and man-made obstacles to enable freedom of movement and maneuver (ATTP 3-90.4). The primary purpose for mobility is to mitigate the effects of natural and man-made obstacles. Mobility operations include reducing, bypassing, or clearing obstacles (including gaps) and marking lanes and trails to enable friendly forces to move and maneuver freely. These tasks frequently occur under conditions that require combat engineer units and most frequently occur when conducted at the tactical level in support of maneuver. Support to early-entry operations includes reconnaissance that would mitigate antiaccess and area-denial mechanisms to clear and open aerial ports of debarkation and seaports of debarkation. These tasks are often considered combat engineering tasks, even though general engineer units can perform them when the conditions allow.

2-11. Engineer tasks to repair, maintain, or build roads, bridges, and airfields usually do not occur under conditions that require combat engineer units. As a result, these tasks are often considered general engineering tasks, even though combat engineer units can perform them, given additional training and augmentation if necessary. Combat engineers can perform these tasks if performed under conditions of close support to maneuver forces that are in close combat.

2-12. Engineer contributions to the planning of mobility operations occur at all levels of war and throughout decisive action. The execution of engineer tasks in support of mobility usually occurs at the operational and tactical levels of war, but will often have strategic level implications. At the tactical level of war, combat engineer units are frequently required, especially in offensive and defensive tasks. At the operational level, engineer tasks are typically performed by general engineer units. During the conduct of offensive and defensive tasks, engineer tasks are focused on the mobility of friendly forces. In stability and DSCA, engineer tasks are often focused on the mobility of the first responders and population.

2-13. Engineer tasks that support mobility operations typically support the assure mobility line of engineer support, but may also support the other three lines. Similarly, a road constructed for an LOC has the purpose of enabling sustainment. Likewise, a bridge might be constructed to develop infrastructure, allowing the local population to transport goods to the market. Engineers perform these tasks most frequently as part of the movement and maneuver warfighting function, but they may perform them in support of the other warfighting functions. Combat engineering is typically focused on mobility at the tactical level, while general engineering is typically focused on mobility at the operational level (although general engineering may impact strategic mobility at times).

2-14. Mobility tasks are typically identified as essential tasks and may require integration into the synchronization matrix to account for the assets and time required to implement them. See chapter 3 for a discussion of planning considerations for mobility, countermobility, and survivability (M/CM/S) operations.

Support to Countermobility

2-15. Engineer support to countermobility includes the following engineer tasks:
- Siting obstacles.
- Constructing, emplacing, or detonating obstacles.
- Marking, reporting, and recording obstacles.
- Maintaining obstacle integration.

2-16. *Countermobility operations* **are those combined arms activities that use or enhance the effects of natural and man-made obstacles to deny an adversary freedom of movement and maneuver.** The primary purpose of countermobility is to slow or divert the enemy, to increase time for target acquisition, and to increase weapon effectiveness. The advent of rapidly emplaced, remotely controlled, networked munitions enables engineers to conduct effective countermobility operations as part of offensive, defensive, and stability tasks and during the transitions among these operations.

2-17. Countermobility tasks typically involve engineers and must always include proper obstacle integration with the maneuver plan, adherence to obstacle emplacement authority, and rigid obstacle control. The engineer advises the commander on how to integrate the obstacle coordinates for the obstacle emplacement authority, establishes obstacle control, recommends directed obstacles, supervises the employment of obstacles, and maintains obstacle status throughout the operation. Most obstacles have the potential to deny the freedom of maneuver to friendly and enemy forces. Therefore, it is critical that the engineer accurately understands the countermobility capabilities and limitations of the available engineer forces and properly weighs the risks of employing various obstacle types. The engineer must also plan for the clearing of obstacles at the cessation of hostilities and for minimizing obstacle effects on noncombatants and the environment.

2-18. The engineer tasks that support countermobility operations include those that construct, emplace, or detonate obstacles and those that track, repair, and protect obstacles. These tasks are often performed by maneuver forces that are in close combat, which require combat engineers units. These conditions frequently occur when the tasks are conducted at the tactical level as part of offensive or defensive tasks. They are often considered combat engineering tasks, even though general engineer units can perform them when the conditions allow.

2-19. The effects of natural and man-made obstacles are considered during planning at every level of war. At the tactical level of war, combat engineers play a prominent role in assessing and predicting the effects and integration of tactical obstacles in support of offensive and defensive tasks. General engineers may also be involved in countermobility operations intended to achieve operational (or strategic) effects. Countermobility operations typically reinforce the terrain to block, fix, turn, or disrupt the enemy ability to move or maneuver, giving the commander opportunities to exploit enemy vulnerabilities or react effectively to enemy actions. In stability operations, countermobility tasks may support missions such as traffic or population control. (See FM 90-7 for information on countermobility.)

2-20. Engineers usually perform these tasks under the first two lines of engineer support—assure mobility and enhance protection—although they may also be applicable in selected cases for the other two lines of engineer support. These tasks typically support the movement and maneuver and protection warfighting functions.

2-21. As of 1 January 2011, U.S. forces are no longer authorized to employ persistent and undetectable land mines (land mines that are not self-destructing or self-deactivating). The current U.S. land mine policy acknowledges the importance of protecting noncombatants while enabling legitimate operational requirements. The United States will continue to employ self-destructing and self-deactivating mines (scatterable mines) to provide countermobility for the force. Additionally, newly developed weapon systems, called *networked munitions*, provide the flexible and adaptive countermobility and survivability capability required by the Army, while conducting unified land operations. Networked munitions are remote-controlled, ground-emplaced weapon systems that provide lethal and nonlethal effects with the ability to be turned on and off from a distance and recovered for multiple employments.

Other Tasks that Assure Mobility

2-22. Geospatial engineering provides the necessary geospatial information and products to help combat and general engineers visualize the terrain and perform tasks along the assure mobility line of engineer support. Geospatial information is the foundation upon which information about the physical environment is referenced to form the common operational picture. (See ATTP 3-34.80 for additional information.) Geospatial information that is timely, accurate, and relevant is a critical enabler throughout the orders process. Geospatial engineers work as staff members to aid in the analysis of the meaning of activities and significantly contribute to the anticipating, estimating, and warning of possible future events. They provide the foundation for developing shared situational understanding, improving the understanding of capabilities and limitations for friendly forces (and the enemy), and highlighting other conditions of the operational environment. Geospatial engineers must possess a thorough understanding of tactics and the application of combat power to tailor geospatial information to support the commander's visualization and decisionmaking. Geospatial engineers provide the following to the assure mobility line of engineer support:

- Three-dimensional perspective fly-through views that enhance terrain visualization within the area of operations, interest, and influence.
- Mobility corridor and combined obstacle overlays to identify assembly areas, plan air and ground missions, and develop engagement areas.
- Fields-of-fire and line-of-sight analysis products to locate defensible terrain, identify potential engagement areas, and position fighting systems to allow mutually supporting fires.
- Urban tactical planners that display key aspects of urban terrain in thematic layers overlaid on high-resolution imagery or maps to facilitate mission planning in urban areas.
- Hydrologic, bathymetric, and gravimetric data analysis to determine soil conditions on land and underwater and the depth of the ocean or lake floors in support of surface and subsurface mobility within the area of operations.
- LOC analysis and overlays to identify structures (roads, airfields, railroads, bridges, tunnels, ferries) capable of facilitating the transportation of people, goods, vehicles, and equipment.

2-23. The engineer diving detachment provides equipment and personnel to conduct underwater operations. The diver's unique skills provide critical support to commanders during river-crossing operations by conducting nearshore and far shore reconnaissance; performing hydrographic surveys to depict bottom composition; conducting underwater and surface reconnaissance of bridges to determine structural integrity and capacity; repairing or reinforcing bridge structures; and emplacing, marking, or reducing underwater obstacles. (See ATTP 3-34.84 and TM 3-34.83 for additional information.)

2-24. Engineer reconnaissance teams may operate independently, but they normally support BCT and regimental combat team reconnaissance squadrons or scout platoons to classify routes, locate obstacles, and determine how to overcome the effects of obstacles by recommending bypass or reduction. Engineer reconnaissance teams also conduct the reconnaissance of proposed obstacle placement locations and ensure that obstacles remain integrated.

2-25. Explosive ordnance clearance agents are combat engineers that are trained to defeat explosive hazards. They can perform limited identification and disposal of unexploded ordnance during route clearance or route reconnaissance, to include blow-in-place, single munitions to enable mobility operations.

ENHANCE PROTECTION

2-26. This line of engineer support is the combination of the three engineer disciplines to support the preservation of the force so that the commander can apply maximum combat power. This line of engineer support consists largely of survivability operations, but can also include selected mobility tasks (construction of perimeter roads), countermobility tasks (emplacement of protective obstacles), and explosive-hazards operations tasks. It also includes survivability and other protection tasks performed or supported by engineers. (See ATP 3-37.34 and ADRP 3-37 for additional information.)

Support to Survivability Operations

2-27. Engineer support to survivability operations consists of the following areas:
- Fighting positions.
- Protective positions.
- Hardened facilities.
- Camouflage and concealment.

2-28. Survivability operations—those military activities that alter the physical environment to provide or improve cover, concealment, and camouflage—are used to enhance survivability when existing terrain features offer insufficient cover and concealment. This is one of the tasks under the protection warfighting function found in ADP 3-37. Engineers employ capabilities from the three engineer disciplines to support survivability operations. Engineer support to survivability operations is most often aligned with the enhance protection line of engineer support.

2-29. Although units conduct survivability operations within capability limits, engineers have a broad range of diverse capabilities that can enhance survivability. Engineer tasks in support of survivability operations include tasks to build, repair, or maintain fighting and protective positions and harden, conceal, or camouflage roads, bridges, airfields, and other structures and facilities. These tasks tend to be equipment intensive and may require the use of equipment timelines to optimize the use of low-density, critical equipment.

2-30. Engineer tasks that support survivability operations occur predominately at the operational and tactical levels of war. At the tactical level of war, they often occur in support to maneuver and special operating forces that are in close combat, which require combat engineer units. This often occurs for tasks to build, repair, or maintain fighting and protective positions. Those tasks are often considered combat engineering tasks, even though general engineer units can perform them when the conditions allow. At the operational level, engineer tasks that support survivability operations are typically performed by general engineer units. In offensive and defensive operations, they are focused on the protection of friendly forces, but during the conduct of stability and DSCA tasks, they sometimes focus on the protection of the population or civilian assets. (See ATP 3-37.34 for additional information about survivability operations.)

Other Tasks

2-31. Engineers also enhance protection through explosive hazards tasks. (See FM 3-34.210 for additional information.) These include area and route clearance; specialized search using engineer mine detection dogs and specialized search dogs; and the collection, analysis, and dissemination of explosive hazards information. These efforts to mitigate the effects of explosive hazards can be performed by engineers at all echelons or by specialized units (explosive-hazards teams, area clearance platoons). (See ATTP 3-34.23 for additional information.)

2-32. Engineers that have trained with EOD personnel and have explosive ordnance clearance experience not only play a vital role in the assure mobility line of engineer support, but are also equally vital for the enhance protection line of engineer support. Explosive ordnance clearance personnel advise the on-scene commander on recommended personnel and equipment protective measures and isolate blast and fragmentation danger areas within the area of operations. Explosive ordnance clearance personnel are trained engineers and may assist EOD personnel in disposing of explosive hazards and with shortages of crucial EOD resources.

Chapter 2

2-33. Engineer mobility and countermobility tasks typically support the assure mobility line of engineer support, but may also support the enhance protection line of engineer support. Constructing a trail for use as a perimeter road to secure a base and providing protective obstacles or entry control points for the protection of base camps are two examples. (See ADP 3-37 for additional information.)

2-34. Engineer divers enhance protection through force protection dives by identifying and removing underwater hazards. Engineer divers improve underwater security measures by checking for the enemy tampering of ships, docks, piers, intakes, and other marine facilities. Engineer divers are trained in explosives and can identify and remove explosive hazards through sympathetic detonation. Planners and senior staffs should be aware of diver capabilities and integrate them into early-entry operations.

2-35. Firefighting teams are limited assets that provide fire prevention and fire protection services. Some of the key protection tasks provided to commanders are fire prevention inspections and investigations, fire suppression, search and rescue, and hazardous material response. Additionally, these teams provide first-level medical response and assistance to victims and technical oversight of nonfirefighting personnel when supporting firefighting operations.

2-36. Other specialized engineer support teams can be embedded at the tactical level to conduct baseline surveys and environmental assessments that enhance protection. These teams identify potential hazards before force projection or base and base camp establishment. (See FM 3-34.5 for additional information on environmental considerations.)

ENABLE FORCE PROJECTION AND LOGISTICS

2-37. Tasks in the enable force projection and logistics line of engineer support free combat engineers to support maneuver forces, establish and maintain the infrastructure necessary to support follow-on forces, and sustain military operations to enable force projection and logistics to continue after hostile action and to provide recommendations for the site selection of facilities, joint fires, and protection. Engineers combine capabilities from the three engineer disciplines to enable force projection and logistics. Primarily through the general engineering discipline, these capabilities are applied to enhance strategic through tactical movements. Tasks in this line of engineer support sustain military operations in-theater.

Tasks That Support Enable Force Projection and Logistics

2-38. The engineer-focused tasks are typically performed by engineer units or commercial contract construction management assets, such as USACE (FFE, Reachback Operations Center), for specialized and reachback support. They can be performed by a combination of joint engineer units, civilian contractors, and HN forces or multinational engineers. They may also require that various types of technical and tactical reconnaissance and assessments be performed before or early in a particular mission. This may include countermobility, site selection, master planning, support to disaster preparedness planning response, and support to consequence management. (See FM 3-34.170 for additional information.)

2-39. Geospatial engineers can provide geospatial products to clarify situational understanding to support operations. This provides early-entry forces with information on soil conditions for landing sites, just as it provides follow-on forces with information on potential locations of bases and base camps for initial operations.

2-40. Combat engineers can provide support that enables force projection and logistics by conducting reconnaissance and clearance tasks. Combat engineers conduct route reconnaissance to determine trafficability and route classification within an area of operations. These engineers also detect and mark explosive hazards and can clear the hazards that are within capability to ensure the freedom of movement along LOCs, aerial ports of debarkation, and seaports of debarkation.

2-41. Engineer personnel augment sustainment units to support joint logistics over the shore to assist planning efforts. Engineer personnel prepare access routes to and from the beach when port facilities are unavailable, damaged, or denied; and they prepare landing sites and staging areas.

Other Tasks That Enable Force Projection and Logistics

2-42. These tasks are primarily general engineering tasks and are not usually performed under conditions of support to maneuver forces that are in close combat. (See FM 3-34.400 for additional information.) These tasks include—
- Constructing and maintaining strategic and operational LOCs, airfields, seaports, railroads, bases and base camps, pipelines, bulk and distribution storage facilities, and standard and nonstandard bridges.
- Providing facilities engineer support.
  - Mobile electrical power.
  - Utilities and waste management.
  - Real estate acquisition, management, remediation, and disposition.
  - Firefighting.
- Conducting battle damage repair.
- Conducting baseline surveys and environmental assessments.
- Improving fighting and protective positions and hardening facilities.

2-43. Prime power is a unique Army capability that provides continuous, reliable, commercial-grade, low- to medium-level voltage to support military operations. It is nontactical power and is typically used to support critical facilities and base camps when commercial power is unavailable or where tactical generators are impractical.

2-44. Real estate teams are a unique Army capability that acquires right-of-way and real estate to enable force projection and logistics support for theater opening and facilitating the transfer and return of real estate back to the HN. General engineers also provide master planning, design, and maintenance for base camps and utilities.

2-45. Engineer divers provide unique capabilities along the enable force projection and logistics line of support. Diver tasks may include the following:
- Neutralize obstacles (that block shipping channels in port and other navigable waterways).
- Repair or reinforce damaged subsurface structures as port facilities, dams, and bridges.
- Conduct search and recovery to locate and salvage submerged equipment, supplies, and personnel.
- Provide support to joint logistics over-the-shore operations by providing a survey of potential beachheads and installing and maintaining offshore mooring systems.

BUILD PARTNER CAPACITY AND DEVELOP INFRASTRUCTURE

2-46. Engineers combine capabilities from across the three disciplines to support the build partner capacity and develop infrastructure line of engineer support, which are vital to stability and counterinsurgency tasks that do not align with a specific phase of operations. This line consists primarily of building, repairing, and maintaining various infrastructure facilities; providing essential services and, ultimately, building partner capacity to codevelop HN capabilities to perform such tasks. Linkages to stability are predominant in this line. Most infrastructure development takes place during Phase 0 (shape), Phase I (deter), Phase IV (stabilize), and Phase V (enable civil authority). It is often a series of technical tasks (build a road, build a water treatment facility) that fall under different sectors (electricity, road and rail transportation, water supply and sanitation, water treatment and sewage).

Tasks That Support Build Partner Capacity and Develop Infrastructure

2-47. This line of engineer support consists primarily of general engineering tasks. Many of the tasks that support this line of engineer support are the general engineering tasks listed previously in the enable logistics line of engineer support. However, the key difference from the enable logistics line of engineer support is the purpose and desired effect. The primary purpose of the tasks in the build partner capacity and develop infrastructure line of engineer support is to support the commander in improving the conditions of

HN leaders, institutions, and infrastructure development capabilities and in influencing them to achieve military objectives for self-defense.

2-48. The different purposes of build partner capacity and develop infrastructure to enable force projection and logistics will significantly change the manner in which the task is executed in most cases. For example, building a road could be a task for the enable force projection and logistics line of engineer support or the build partner capacity and develop infrastructure line of engineer support. While the completed road may be the same, the conditions and requirements to build it may be very different due to its intended purpose. If the road is being built to improve the local economic conditions, using local labor to increase employment may be more important than just completing the work in the quickest manner possible. Additionally, a road for the local populace may require coordination with many different local agencies, organizations, and ministries to support the local government and assist them in establishing legitimacy. Engineers may be required to provide technical training to HN managers on engineer tasks in planning, designing, and constructing roads. The interaction with the population in the process of building the road may likely take priority over the quality and speed of completion of the road itself.

2-49. Included in the build partner capacity and develop infrastructure line of engineer support is the engineer role in capacity building. (See FM 3-07 for additional information.) Engineers may support the U.S. Agency for International Development, the State Department, and special operations forces to improve HN infrastructure and the human or intellectual capacity to sustain sector over time. Tasks to improve HN infrastructure will require coordination with local- or national-level government agencies or ministries that maintain or control infrastructure. The tasks may emphasize the development of local technical and engineering institutions. Engineers may be required to train, educate, and develop local leaders, engineers, and organizations in the process of executing a task in this line of engineer support. For example, an engineer unit that is assisting the local populace in improving drinking-water systems will also train the local public works to operate and maintain the system.

2-50. While engineers at all echelons build partner capacity requirements, USACE FFE units have additional expertise to advise and assist HN capacity building that spurs long-term relationships. Engineers supporting BCTs may build partner capacity by providing training teams and reconstruction teams, sharing institutional knowledge, and conducting key leader engagements.

Other Tasks That Build Partner Capacity and Develop Infrastructure

2-51. General and geospatial engineers contribute to this line of engineer support in that geospatial engineers and other USACE experts can provide technical advice and assistance. Specialized units can locate and map water sources. Well-drilling teams are limited assets that can be applied to solve long-term water restoration issues.

2-52. Engineers across the disciplines may support tasks that build partner capacity and develop infrastructure by participating in foreign exchange programs and attending conferences. Participation in joint exercises is another opportunity that allows engineers to exchange information, build relationships, and develop infrastructure simultaneously.

ENGINEER SUPPORT TO WARFIGHTING FUNCTIONS

2-53. Unified land operations require the continuous generation and application of combat power, often for protracted periods. *Combat power* is the total means of destructive, constructive, and information capabilities that a military unit and formation can apply at a given time (ADRP 3-0). Army forces generate combat power by converting potential into effective action. (See ADRP 3-0 for additional information.) There are eight elements of combat power—leadership, information, mission command, movement and maneuver, intelligence, fires, sustainment, and protection. Leadership and information also multiply the effects of the other six elements of combat power. These six—mission command, movement and maneuver, intelligence, fires, sustainment, and protection—are collectively described as the warfighting functions. In unified land operations, Army forces combine the elements of combat power to defeat the enemy and master each situation.

2-54. Engineers provide support not only to the six warfighting functions, but also to the special operations forces community. Engineers support special operations in the same manner in which they support

Engineer Support to Unified Land Operations

warfighting functions. The engineer disciplines are well suited to provide engineer support for special operations from an advisory role to augmentation support. Engineering capabilities are scalable and can be tailored to provide horizontal and vertical construction capabilities to improve austere conditions. As special operations forces tend to deploy into smaller formations, engineers will often provide support through the supervision of HN contractors and laborers.

2-55. Engineer support contributes significant combat power (lethal and nonlethal) to unified land operations. To support the combined arms team effectively, engineering capabilities are organized by the engineer disciplines and synchronized in the application through the warfighting functions. These warfighting functions also provide the framework for engineer tasks in the Army universal task list.

2-56. Every unit, regardless of type, generates combat power and contributes to the operation. A variety of engineering capabilities and unit types are available to contribute to combat power. Engineer disciplines are each generally aligned in support of specific warfighting functions, although they have impact in and across the others. (See figure 2-1.) For example—

- Survivability support may be provided with linkages to the fires warfighting function.
- Combat engineering is aligned primarily with the movement and maneuver and protection warfighting functions.
- General engineering aligns to focus its support on the sustainment and protection warfighting functions and the reinforcement of combat engineering outside close combat.
- Geospatial engineering is primarily aligned with the mission command and intelligence warfighting functions.

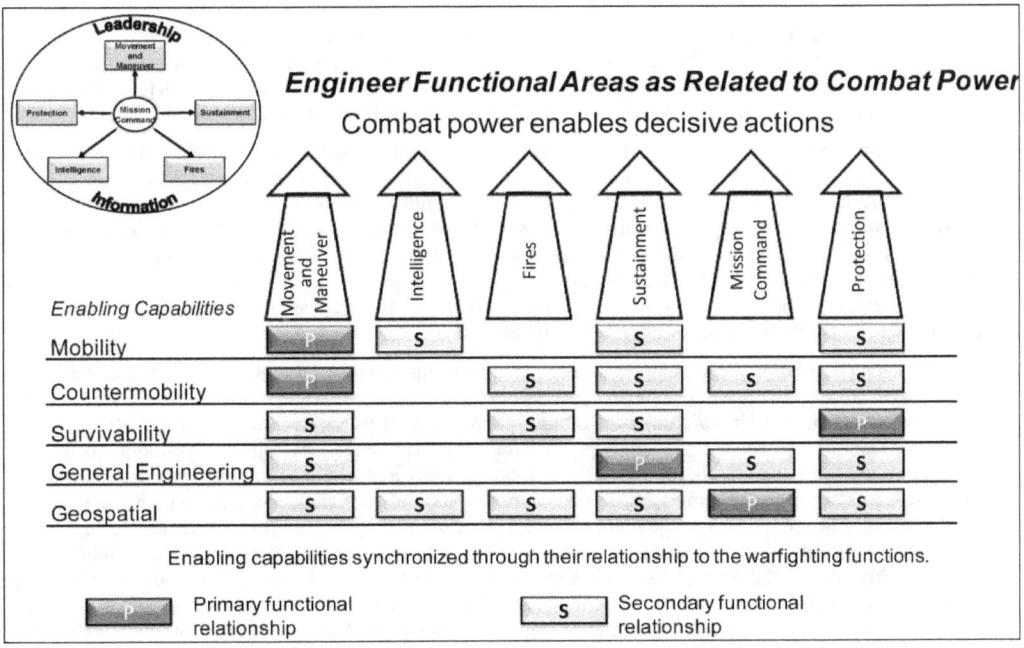

Figure 2-1. Engineer application of combat power

2-57. *Combined arms* is the synchronized and simultaneous application of arms to achieve an effect greater than if each arm was used separately or sequentially (ADRP 3-0). The warfighting functions provide engineers with a common framework to link the required engineering capabilities to the synchronized application of combined arms.

MISSION COMMAND

2-58. The *mission command warfighting function* is the related tasks and systems that develop and integrate those activities enabling a commander to balance the art of command and the science of control in order to integrate the other warfighting functions (ADRP 3-0). It is unique in that it integrates the activities of the other warfighting functions.

2-59. Engineer units must integrate mission command and the operations process activities for the unit while interacting with the mission command activities of the unit being supported. The interaction may be primarily through an engineer staff assigned to the supported unit or through staff counterparts. In some cases, a supported unit may not have assigned engineer staff and the supporting unit will provide this support as well. This relationship and degree of interaction is determined by many factors, including the type of unit and echelon being supported and the command or support relationship established. This manual addresses mission command of engineer forces separately from engineer staff participation in the supported commander mission command.

2-60. There are typically not enough engineering capabilities available to accomplish the desired engineer tasks. Careful prioritization must occur. Even more challenging is that once in the area of operations, force-tailored engineer units must be able to rapidly transition among elements of operations.

2-61. Because the available force-tailored engineer units are designed for specific tasks, engineering capabilities must be shifted within the area of operations to match the requirements with the capabilities of the modular engineer units. Transitions will occur at the strategic, operational, and tactical levels; and flexibility in the task organization will be required to permit the shifting of engineering capabilities.

2-62. Control measures are essential tools to help engineers accomplish the mission. One such control measure is the engineer work line, which is a graphic control measure used to designate areas of work responsibility for subordinate engineer organizations. **An *engineer work line* is a coordinated boundary or phase line used to compartmentalize an area of operations to indicate where specific engineer units have primary responsibility for the engineer effort.** The engineer work line may be used at the division level to discriminate between an area of operations supported by division engineer assets and an area of operations supported by direct or general support corps engineer units. (See FM 3-34.400 for additional information on general engineering operations.)

2-63. Whether a subordinate or supporting unit, engineer unit commanders must understand and exercise the art and science of mission command as described in ADP 3-0. (See ADP 5-0 and ADP 6-0 for additional information.) Organic units operating within assigned BCTs operate within that structure as a matter of routine. However, the augmenting units face challenges by quickly task-organizing and integrating into the receiving unit. Similarly, as modular units and headquarters elements are allocated to division, corps, and theater armies, those unit commanders and staff must integrate within the receiving headquarters. The engineer headquarters provides control of ongoing engineer operations, to include monitoring engineer forces and assets, mitigating explosive hazards, coordinating engineer reconnaissance, and providing geospatial support through geospatial information and services. This adds depth to the engineer staff capabilities within the supported or gaining headquarters. Similarly, task-organized units face challenges in quickly recognizing and integrating into the distinct character of the new unit. A thorough understanding of, and practice with, the mission command function and the operations process that it drives enable the flexibility necessary for modular engineer forces to integrate into supported units. In unique cases, where an engineer headquarters serves as the foundation around which a task force or JTF is formed (a disaster relief operation), it is critical for the mission command function and the operations process it drives to adhere closely to the ideal described in Army and applicable joint doctrine.

2-64. Finding ways to accomplish the mission with an appropriate mix of lethal and nonlethal actions is a paramount consideration for every Army commander. Through synchronization, commanders mass the lethal and nonlethal effects of combat power at the decisive place and time to overwhelm an enemy or dominate the situation. Engineer leaders and staff planners at each echelon play a pivotal role in ensuring the synchronization of a variety of engineering capabilities that are available to conduct or support unified land operations.

MOVEMENT AND MANEUVER

2-65. The *movement and maneuver warfighting function* is the related tasks and systems that move forces to achieve a position of advantage in relation to the enemy and other threats (ADRP 3-0). Engineers support the movement and maneuver warfighting function by performing tasks associated with geospatial engineering, engineer reconnaissance, and M/CM/S operations. The three engineer disciplines support the movement and maneuver warfighting function. Combat engineer support applied through the movement and maneuver warfighting function is focused on assured mobility because combat engineers are trained and equipped to support forces in close combat. A BCT organic engineer unit shapes the battlefield to support early-entry operations with mobility and countermobility tasks, which enables initial lodgments and further expands lodgments to enable force projection.

Performing as Combat Engineers

2-66. Operating in close combat support to maneuver forces requires that combat engineer units be able to integrate and coordinate actions with the fire, movement, or other actions of combat forces. To do that, combat engineer units must be organized, manned, equipped, and trained differently than general engineer units who are not designed to operate in combat conditions. For example, combat engineer units are organized similarly to infantry squads and platoons, manned with additional medical personnel, equipped with specific weapons and vehicles, and trained with supported close combat force. These requirements limit the ability of combat engineer units to perform many tasks to the same standard as general engineering units. Combat engineer units can perform as general engineers, and vice versa, with additional equipment, training, and augmented technical expertise.

Fighting as Engineers

2-67. Fighting as engineers is inherent to the primary mission of engineer units. Combat engineers are well forward because they fight alongside maneuver units as part of a combined arms team. When supporting unified land operations, engineers must be prepared to fight and employ combat skills and integrate activities with fire and maneuver. On the battlefield, the enemy will make every effort to detect and engage engineers quickly, regardless of location. In addition to the primary responsibilities within combat engineering, combat engineers are trained, organized, and equipped to fight and destroy the enemy. Combat engineers engage in close combat to accomplish engineer missions and to—

- Neutralize explosive hazards by locating, assessing, and rendering them incapable of interfering with the conduct of operations (except render-safe procedures).
- Enhance mobility by conducting route and obstacle reconnaissance, obstacle reduction, assault gap crossing, construction and repair of combat road and trails, and forward aviation combat engineering.
- Deny the enemy the freedom of movement and maneuver (countermobility) by lethal and nonlethal means with land mines, network munitions, and demolition and constructed obstacles.
- Enhance protection through survivability operations (fighting and protective positions, hardening facilities, camouflage and concealment).

Fighting as Infantry

2-68. Throughout history, engineer organizations have been required to fight as infantry as a secondary mission. A combat engineer organization is capable of executing infantry tasks or task-organizing to fight as infantry with other combat units. When reorganized, combat engineers require additional positions normally found in maneuver formations (fire control, medical personnel). If an engineer battalion has been designated to reorganize and fight as infantry, it requires the same support and integration as maneuver units (armor, fire support) in its task organization to accomplish the mission. It may also require significant reorganization. A commander who commands combat engineers has the authority to reorganize them as infantry, unless otherwise reserved. However, a commander must carefully weigh the gain in infantry strength against the loss of engineer support.

2-69. Reorganizing engineer units as infantry requires careful consideration and is normally reserved to the operational level command. Reorganization involves extensive equipment and training that are specific to the reorganization, and it must be coordinated with the higher headquarters. Employing engineers merely implies that the gaining commander will be using the engineers for a short period. Reorganization also requires resources, time, and training.

2-70. An emergency or immediate requirement for infantry may not require the reorganization of engineers. Engineers may simply be required to engage in close combat. Commanders should consider this option in limited scope and task application. The commander makes a decision after weighing the mission variables; determining an acceptable risk level; and considering the resources, time, and training required to reorganize engineer units as infantry.

2-71. General engineer support to movement and maneuver accomplishes the tasks that exceed the capability of the combat engineer force. General engineer support to movement and maneuver also accomplishes extensive upgrades or new construction of LOCs and base camps. (See FM 3-34.400 for additional information.) Although general engineer support is typically applied through the sustainment warfighting function, it may include many of the following tasks that also cross over to support movement and maneuver:

- Constructing and repairing combat roads and trails that exceed the capability of combat engineer assets.
- Providing forward aviation combat engineering that exceeds the capabilities of combat engineer assets.
  - Repairing paved, asphalt, and concrete runways and airfields.
  - Conducting airfield surveys.
  - Providing firefighting and aircraft rescue services.
  - Marking airfield landing and parking surfaces.
- Constructing standard and nonstandard bridging.
- Ensuring theater access through the construction and upgrade of LOCs, main supply routes, ports, airfields, and base camps.

2-72. Engineer units may be called on to provide assets to contribute to maneuver support operations when assigned to a MEB. (See FM 3-90.31 for additional information.) Missions assigned to engineers in the conduct of maneuver support operations will enable one or more key tasks related to MEB primary missions. Below are a few tasks that are generally associated with engineers:

- Conducting M/CM/S operations.
- Performing area damage control.
- Restoring essential services.

INTELLIGENCE

2-73. The *intelligence warfighting function* is the related tasks and systems that facilitate understanding the enemy, terrain, and civil considerations (ADRP 3-0). Engineering capabilities can be employed to add to the situational understanding of the commander. Engineers play a major role in the intelligence preparation of the battlefield process by anticipating and providing digitized mapping and terrain analysis products of likely contingency areas. Geospatial engineering improves terrain visualization and understanding of the physical environment. It is an essential contributor to geospatial intelligence. Engineer staff and planners provide a predictive and deductive analysis of enemy engineering capabilities to intelligence.

2-74. Engineer reconnaissance provides data and information that contributes to answering the commander's critical information requirements and is necessary in the lines of engineer support. (See FM 3-34.170 for additional information.) To accomplish all four lines of engineer support, engineers must designate the specialized assets available to collect the information needed to answer these requirements. Reconnaissance is inherent in the three disciplines; however, the information collected may be different and tactical or technical in nature. The engineer disciplines provide a menu of reconnaissance capabilities.

These vary in linkages to warfighting function tasks. They also vary in the type and degree of tactical or technical expertise and effort. The capabilities are provided and organized by combat and general engineer units with overarching support from geospatial means. These units do not have organized and dedicated reconnaissance elements within the structure (except for the armor brigade combat team) but are organized with a mix of engineer specialties, expertise, and equipment. Commanders task-organize combat and general engineers with other elements from across the engineer disciplines or warfighting functions based on the mission and situation.

2-75. Reconnaissance in support of M/CM/S operations is conducted primarily by engineer reconnaissance teams. Engineer reconnaissance teams are composed of combat engineers and are focused on the collection of tactical and technical information to support the BCT freedom of maneuver and survivability of friendly forces and facilities. With the exception of the armor BCT, the current engineer force structure does not provide for personnel or equipment dedicated to reconnaissance efforts. This requires the engineer company commanders to form and train ad hoc teams for tactical reconnaissance tasks that focus on collecting technical information and performing a limited analysis.

2-76. Engineer information collection is a deliberate process. The engineer information collected assists commanders in determining the feasibility of areas for use based on the aspects of the terrain. Engineer information collection may be conducted remotely or physically, but it is an essential task performed during planning. An assessment of the area of operations begins well before the deployment of forces, and continuous assessments ensure that accurate information is provided to the common operational picture. Engineer information collection may include, but is not limited to, conditions and capacities that support mobility, potential sources of construction materials, local construction standards, power generation and transmission capabilities, and geotechnical data in the area of operations (soils, geology, hydrography). Engineer staffs at division, corps, theater army echelon, and in-theater engineer headquarters determine engineer information requirements in an area of operations; and they collect and analyze engineer information in coordination with the respective assistant chief of staff, intelligence.

2-77. Engineer intelligence is a deliberate process. Data collection assists commanders with engineering and sustainment operations. Engineer intelligence preparation of potential areas of operations begins well before the deployment of forces. Engineer intelligence may include, but is not limited to, conditions and capacities of ports of debarkation, potential sources of construction materials, local construction standards, power generation and transmission capabilities, and geotechnical data in the area of operations (soils, geology, hydrography). Engineer staffs at division, corps, theater army echelon, and in-theater engineer headquarters determine engineer intelligence requirements in a potential area of operations, and they collect and analyze engineer intelligence data in coordination with the respective assistant chief of staff, intelligence.

2-78. Geospatial engineering teams apply information and services to improve the situational understanding of terrain. *Geospatial information and services* is the collection, information extraction, storage, dissemination, and exploitation of geodetic, geomagnetic, imagery, gravimetric, aeronautical, topographic, hydrographic, littoral, cultural, and toponymic data accurately referenced to a precise location on the Earth's surface (JP 2-03). The Army Geospatial Center is a reachback capability that includes instruction, training, and guidance for the use of geospatial data to enable users to access and manipulate data. Common military applications of geospatial information and services include support to—

- Planning.
- Training.
- Geospatial intelligence and operations.
    - Navigation.
    - Mission planning.
    - Mission rehearsal.
    - Modeling.
    - Simulation.
    - Precise targeting.

Chapter 2

FIRES

2-79. The *fires warfighting function* is the related tasks and systems that provide collective and coordinated use of Army indirect fires, air and missile defense, and joint fires through the targeting process (ADRP 3-0). Engineering capabilities significantly contribute to this warfighting function when they are used to facilitate targeting. Geospatial engineers may provide template observer and firing points based on the line of sight and the slope restrictions and may analyze the mobility and suitability of potential targets and engagement areas to facilitate the repositioning of artillery systems. Combat engineers may be used to shape terrain by emplacing obstacles that enhance the effect of fires, construct survivability positions for fires units, and support mobility during displacements.

SUSTAINMENT

2-80. The *sustainment warfighting function* is the related tasks and systems that provide support and services to ensure the freedom of action, extend operational reach, and prolong endurance (ADRP 3-0). Engineers support the sustainment warfighting function by performing tasks associated with mobility operations and survivability operations. Engineers contribute by constructing base camps, ammunition holding areas, and revetments or other types of hardening of distribution facilities and by clearing LOCs.

2-81. General engineer applications are primarily linked through a major category of tasks that provide logistics support in the sustainment warfighting function. As already discussed, general engineering capabilities can also be applied in support of combat engineer applications and will have links across the movement and maneuver warfighting function and protection warfighting function.

2-82. During the conduct of stability or DSCA tasks, sustainment support may shift to the establishment of services that support civilian agencies and the normal support of U.S. forces. The conduct of stability tasks tends to be of a long duration compared to the other operations. As such, the general engineering level of effort, including FFE support from USACE, is very high at the onset and gradually decreases as the theater matures. As the area of operations matures, the general engineering effort may transfer to theater or external support contracts (logistics civil augmentation program, Air Force contract augmentation program, Navy global contingency construction contract).

2-83. Contracting support obtains and provides supplies, services, and construction labor and materiel—often providing a responsive option or enhancement to support the force. (See FM 4-92 and ATP 4-94 for additional information.) General engineers will often be required to provide subject matter expertise for the supervision of contracted services and materials use.

PROTECTION

2-84. The *protection warfighting function* is the related tasks and systems that preserve the force so the commander can apply maximum combat power to accomplish the mission (ADRP 3-0). Engineers have unique equipment and personnel capabilities that can be used to support survivability operations and related protection tasks. Combat engineers, supported by general engineer capabilities when required, provide selected survivability operations through the protection warfighting function. (See ATP 3-37.34 for additional information.) Combat engineers typically provide the basic hardening and field fortification support, while general engineer support is focused on long-term survivability efforts. General engineer support is also applied through the protection warfighting function to control pollution and hazardous material and to harden facilities. Survivability operations include the following engineer tasks:

- Protecting against enemy hazards within the area of operations.
  - Constructing vehicle fighting positions, crew-served weapon fighting positions, or individual fighting positions.
  - Constructing protective earth walls, berms, and revetments or constructing vehicle, information systems, equipment, and material protective positions.
  - Employing protective equipment (vehicle crash barriers, entry control points, security fences).
  - Installing bridge protective devices for an existing float bridge or river-crossing site to protect against waterborne demolition teams, floating mines, or floating debris.

- Installing or removing protective obstacles.
- Conducting environmental assessments to identify and protect against environmental conditions.
* Conducting actions to control pollution and hazardous material. (See FM 3-34.5 for additional information.)
* Conducting tactical firefighting. (See ADRP 3-37 and FM 5-415 for additional information.)

2-85. When conducting stability or DSCA tasks, survivability remains a key concern. Although the likelihood of combat operations is reduced, key resources and personnel remain vulnerable to other types of hostile action or attack. Commanders must consider protecting vital resources such as fuel sites, sustainment convoys, base camps, and logistics support areas since the entire area of operations has an equal potential for enemy attack. Therefore, the priority of work for construction assets will be much more focused on protecting these types of resources than on constructing fighting positions for combat vehicles or crew-served weapons. Vital resources requiring survivability may also include facilities critical to the civilian infrastructure (such as key industrial sites, pipelines, water treatment plants, and government buildings). Engineers also employ protective obstacles as a key tool in protecting these important assets and locations. Protective obstacles range from tetrahedrons and concrete barriers to networked munitions. Physical barriers provide relatively inexpensive, inflexible survivability capability. Networked munitions, with built-in sensor capabilities and central control, provide a flexible intrusion detection and denial system.

TASKS SUPPORTING DECISIVE ACTION

2-86. Decisive action requires simultaneous combinations of offense, defense, and stability or DSCA tasks. ADRP 3-0 lists the tasks associated with each element and the purposes of each task. Each task has numerous associated subordinate tasks. Engineering capabilities are organized by the engineer disciplines and are synchronized in the application through the warfighting functions. As described in chapter 3, the operations process activities provide the context in which the synchronization and application are integrated into the combined arms operation.

OFFENSIVE TASKS

2-87. Engineer support to the offense includes the simultaneous application of combat, general, and geospatial engineering disciplines through synchronizing warfighting functions and throughout the depth of the area of operations. Combat engineering in support of maneuver forces is the primary focus of engineers involved in the conduct of offensive tasks; however, the three disciplines are applied simultaneously to some degree. The primary focus will be support that enables movement and maneuver. Figure 2-2, page 2-16, shows a notional application of engineering capabilities supporting offensive operations.

2-88. Combat engineers use preparation activities to posture engineer assets with the task-organized gaining or supported headquarters. Engineer units establish early linkups with the maneuver units they will support. As combat engineer units prepare for offensive tasks, they focus on inspections and combined arms rehearsals. Combined arms breaching and gap-crossing forces are task-organized, and they conduct rehearsals for the breach, assault, and support forces. The engineer staff officer at the appropriate echelon coordinates engineer reconnaissance that is focused to support the collection of the appropriate information to create obstacle intelligence. Tactical bridging equipment that is used to create combat trails or forward airfields and landing zones is moved into staging areas. If route clearance operations are anticipated, clearance teams are task-organized and focused on inspections and combined arms rehearsals. Combat engineer preparation activities occur in close proximity and are closely aligned and integrated with maneuver force preparations.

Chapter 2

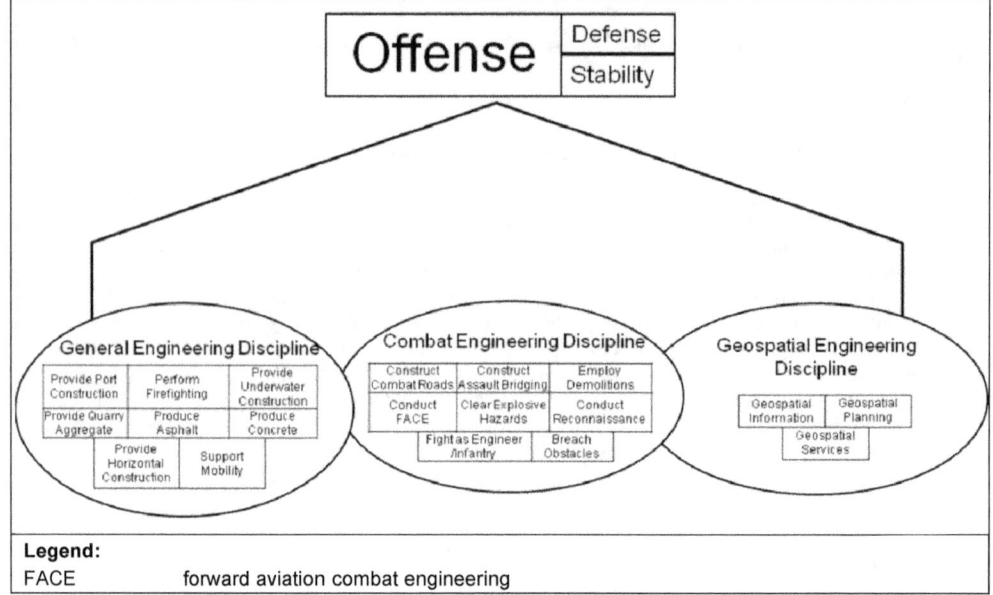

Figure 2-2. Notional engineer support to offensive tasks

2-89. Engineer staff officers at every echelon coordinate the movement and positioning of general engineer assets that are task-organized to augment combat engineering capabilities. Although general engineer assets can be placed in command or support relationships with the maneuver force, a command relationship with the supported engineer unit is often more effective. General engineer assets will require more time for movement given the nature of the heavy, wheeled equipment that is employed. For significant construction, preparation activities may require a more technical engineer reconnaissance to enable adequate project planning and design, including the provision of construction materials as required. Specialized engineer assets may also be necessary to accomplish certain missions. At the operational level, general engineer activities may not be conducted as part of a combined arms mission, but must be fully coordinated with the maneuver commander responsible for the area of operations. Such general engineer support is primarily applied to enable the sustainment warfighting function, but may also be critical to the preparation for an offensive operation, to include support to operational mobility.

2-90. During the conduct of offensive tasks, fighting and protective position development is minimal for tactical vehicles and weapons systems. The emphasis lies on the mobility of the force. Protective positions for artillery, air and missile defense, and logistics positions may be required in the offense and defense, although more so in the defense. Stationary command facilities require protection to lessen vulnerability. During halts in the advance, the terrain will provide a measure of protection. Therefore, units should develop as many protective positions as possible for key weapon systems, command nodes, and critical supplies based on the threat level and unit vulnerabilities. For example, expedient earth excavations or parapets are located to make the best use of existing terrain. During the early planning stages, geospatial engineering teams can provide information on soil conditions, vegetative concealment, and terrain masking along march routes to facilitate survivability for the force. Each position design should consider camouflage from the start and the development of deception techniques as the situation and time permit.

2-91. When executing offensive tasks, the maneuver force uses its common operational picture to link detection efforts to maneuver to avoid encountering obstacles along the route of the attack. The maneuver force can actively avoid obstacles by interdicting threat countermobility efforts before emplacement or passively avoiding obstacles by identifying, marking, and bypassing them. Assessments by on-site engineers assist in the decision to bypass or breach obstacles. If the friendly force commander is compelled to neutralize obstacles, the force employs the breach tenets of intelligence, breach fundamentals, breach organization, mass, and synchronization. Bypasses are preferred when possible, and they may be handed off to follow-on engineer units for maintenance and improvement. Similarly, tactical bridging must be replaced when feasible with appropriate LOC or bridging so that assault-bridging assets can remain postured for future missions. Assessments that are more technical are made as soon as possible to determine feasible and suitable improvements to the LOCs.

DEFENSIVE TASKS

2-92. Engineer support to the defense includes the simultaneous application of combat, general, and geospatial engineering capabilities through synchronizing warfighting functions throughout the depth of the area of operations. Combat engineering in close support of maneuver forces is the primary focus in defensive operations; however, the three disciplines are applied simultaneously to some degree. Figure 2-3 shows a notional application of engineering capabilities supporting defensive operations.

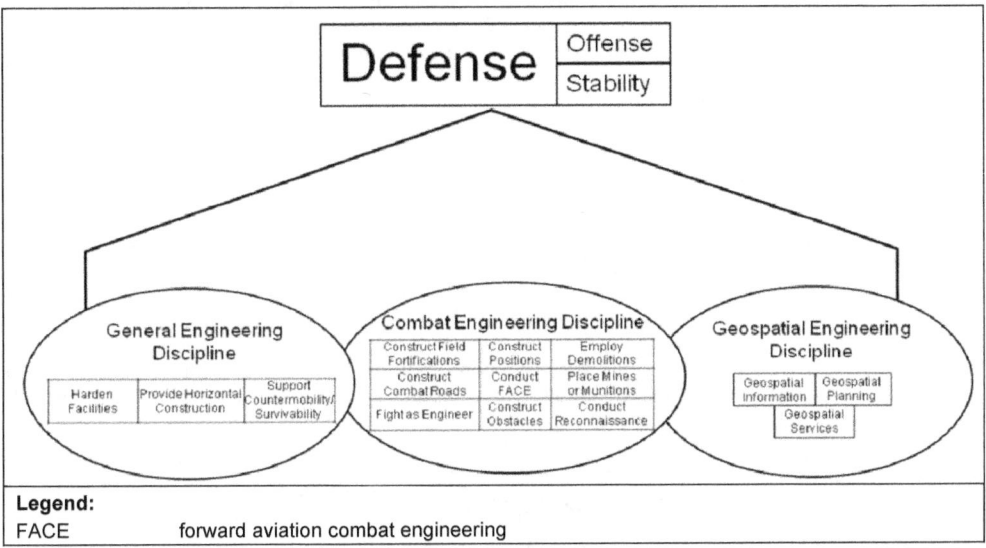

Figure 2-3. Notional engineer support to defensive tasks

2-93. The primary focus for combat engineers in support of defensive tasks is to enable combined arms obstacle integration (countermobility) to facilitate mobility for friendly repositioning or counterattacking forces. Defensive missions demand the greatest survivability effort. Activities in the defense include constructing survivability positions for headquarters, artillery, air and missile defense, and critical equipment and supplies. Activities also include preparing individual and crew-served fighting positions and defilade fighting positions for combat vehicles. The use of engineer work timelines is essential, and digging assets are intensively managed. During this period, countermobility efforts will compete with survivability resources and assets. Because of this, it is critical that maneuver commanders provide clear guidance on resources and priorities of effort. General engineers support tasks that exceed the capability of the combat engineer force and provide more extensive support to the mobility of repositioning counterattack forces. Examples of expected missions include the—
- Construction and integration of obstacles.
- Preparation of fighting positions and survivability positions in depth.
- Upgrade and repair of routes that facilitate the repositioning of forces throughout the area of operations.

2-94. During preparation, engineer assets are postured with the task-organized gaining or supported headquarters, and they initiate the engineer work effort. The equipment work effort is a balance between countermobility and survivability as determined by the commander. The effort continues throughout preparation activities until it is complete or until it is no longer feasible. Significant coordination is required to resource the materials required for constructing obstacles and fighting positions and to integrate the obstacles with friendly fire effects. Designated combat engineers provide mobility support for the reserve or mobile strike force. The engineer staff officer, at appropriate echelons, coordinates for engineer reconnaissance and surveillance assets to identify specific enemy engineering capabilities (breaching, bridging, and countermobility assets) to nominate those capabilities for targeting, ensuring timely destruction.

2-95. At the operational level, general engineer support will be continuously conducted to harden and prepare protective positions for facilities and installations. These activities are primarily applied through the protection warfighting function. General engineering support to protection and survivability continues throughout operations as improvements are continuously reassessed and an additional effort is made available. Operational level obstacles may also be necessary as part of countermobility support (see JP 3-15 and GTA 90-01-011). Other general engineer activities applied to enable the sustainment warfighting function may also be critical to the preparation and conduct of defensive operations. Enabling mobility throughout the depth of the area of operations will remain an engineer mission.

STABILITY TASKS

2-96. Stability operations consist of five primary tasks—civil security, civil control, essential services restoration, support to governance, and support to economic and infrastructure development. The primary tasks are discussed in detail in ADRP 3-07.

2-97. Engineer support to stability operations includes the simultaneous application of combat, general, and geospatial engineering capabilities through synchronizing warfighting functions and throughout the depth of the area of operations. General engineering support for the restoration of essential services and infrastructure development is the primary engineer focus in stability operations; however, the three disciplines are applied simultaneously to some degree. Figure 2-4 shows a notional application of engineering capabilities providing support to stability operations. The participation of engineer generating-force elements (USACE tasks provided during stability tasks) will be significant and is typically realized as general or geospatial engineering support.

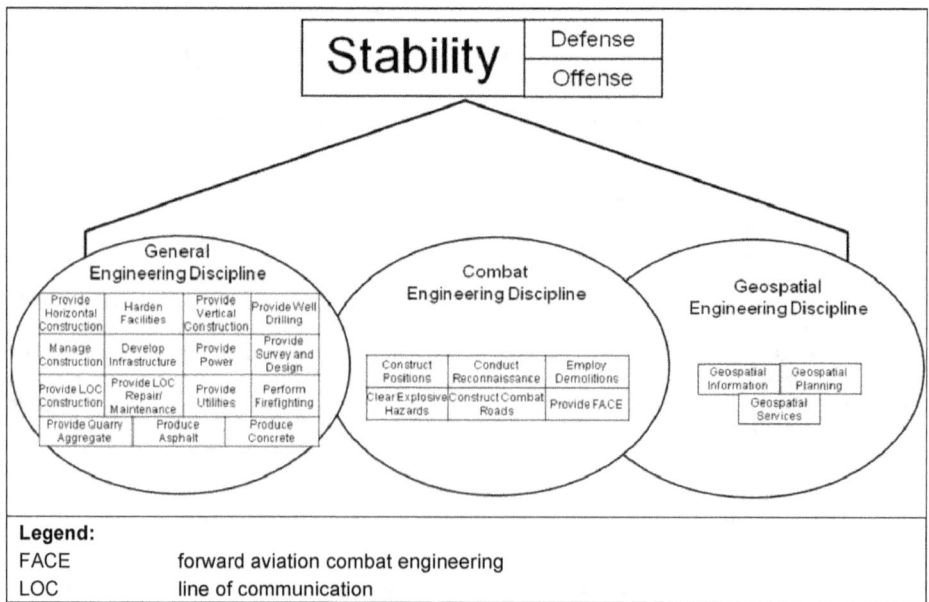

Figure 2-4. Notional engineer support to stability tasks

2-98. Often, stability operations are required to meet the critical needs of the populace. Engineer forces may be critical enablers in the provision of essential services until the HN government or other agencies can provide essential services. Engineer tasks primarily focus on establishing or reconstructing infrastructure to provide essential services that support the population. The effort is typically conducted in conjunction with civilian agencies and other engineer support of U.S. forces. The support for infrastructure development may be extended to assist the HN in developing capability and capacity. Essential services for engineer consideration include food and water, emergency shelter, and basic sanitation (sewage and waste disposal). Likely engineer stability tasks are similar to those required in DSCA, except that they are conducted overseas. These tasks include the following:

- Constructing and repairing rudimentary surface transportation systems, basic sanitation facilities, and rudimentary public facilities and utilities.
- Detecting and assessing water sources and drilling water wells.
- Constructing feeding centers.
- Providing environmental assessment and technical advice.
- Constructing waste treatment and disposal facilities.
- Providing base and base camp construction and power generation.
- Conducting infrastructure reconnaissance, technical assistance, and damage assessment.
- Conducting emergency demolitions.
- Conducting debris or route clearing operations.

2-99. Engineer support to stability tasks may include the typical integration with, and support for, combined arms forces in their missions. Combat engineer route clearance and other close support capabilities may be critical tasks that are applied through the movement and maneuver warfighting function. Geospatial engineer support continues to provide foundational information that supports the common operational picture. General engineer support may be required for the sustainment and protection requirements of the force. However, in stability operations, a focus of the engineer effort is likely to be the general engineering capabilities applied to restore essential services and support infrastructure development.

2-100. Many of the technical capabilities only found in the generating force will be essential to providing appropriate engineer support as those elements of the Engineer Regiment are called upon (through reachback and FFE) for specialized expertise and capabilities. Stability operations tend to be of a long duration compared to the other unified land operations. As such, the general engineering level of effort is very high at the onset and it gradually decreases as the theater matures; support will be required to some degree for the duration of the stability operation. Preparation activities include the identification of significant infrastructure and base development construction projects and the nomination of those projects for funding. The highest priority projects may be executed using military general engineer capabilities, while others may compete for contingency funding and execution through a contract capability. As the area of operations matures, the general engineering effort in support of sustainment requirements may transfer to theater or external support contracts (logistics civil augmentation program, Air Force contract augmentation program, Navy global contingency construction contract).

2-101. Engineer support may be critical to civil affairs operations, which may include activities of nonmilitary organizations and military forces. Similarly, engineering capabilities may be applied to provide specific construction and other technical support integrated within the commander's plan. Integration occurs throughout the operations process, and it is facilitated by coordination between the engineer staff officer and civil affairs staff at the civil-military operations center.

2-102. Preparing for stability operations may be more difficult than preparing for combat operations because of the technical nature of the requirements and the broad range of potential engineer missions associated with them. An early, on-the-ground assessment can be critical to tailor the engineer force with required specialties and engineer resources. The results of this assessment are passed to planners to ensure that an adequate engineer force arrives in the area of operations in a timely manner. This early, on-the-ground engineer reconnaissance and associated assessment or survey identify the—

- Status of the infrastructure in the area of operations (airfields, roads, ports, logistics bases, troop bed-down facilities); real estate acquisition; environmental standards, conditions, and considerations; construction material supply; construction management; and line-haul requirements.
- Status of theater- and situation-specific protection requirements.
- Availability of existing geospatial products and requirements for new terrain visualization products.
- Requirements for specialized engineer support (prime power, well drilling, quarry, firefighting) and support to other emergency services.
- Status of specialized engineer requirements available only in the capabilities of generating-force elements of the Engineer Regiment.
- Requirements of the mission command system, to include headquarters staffing, communications, and information systems support.
- Requirements for engineer liaison, to include linguists and civil affairs personnel.
- Potential for contract construction or other engineering capabilities.

DEFENSE SUPPORT OF CIVIL AUTHORITIES

2-103. DSCA includes operations that address the consequences of natural or man-made disasters, accidents, and incidents within the United States and its territories. Army forces conduct DSCA when the size and scope of events exceed the capabilities or capacities of domestic civilian agencies. The Army National Guard is often the first military force to respond on behalf of state authorities. DSCA includes four primary tasks (ADRP 3-28 discusses these tasks in detail)—

- Provide support for domestic disasters.
- Provide support for domestic chemical, biological, radiological, and nuclear incidents.
- Provide support for domestic civilian law enforcement agencies.
- Provide other designated support.

2-104. Engineering in DSCA may include the simultaneous application of combat, general, and geospatial engineering capabilities through synchronizing the warfighting functions throughout the area of operations. General engineering support for the restoration of essential services is the primary engineer focus in DSCA. Engineer support may also be required for Army forces providing mission command, protection, and sustainment to government agencies until they can function normally. Figure 2-5 shows a notional application of engineering capabilities supporting DSCA. The generating-force elements of the Engineer Regiment, such as the USACE, will play a critical and significant role in DSCA.

2-105. There are few unique engineer missions performed in DSCA that are not performed during other operations. The difference is the context in which they are performed. The U.S. law carefully limits the actions that military forces, particularly Regular Army units, can conduct within the United States and its territories. In addition to legal differences, DSCA is always conducted in support of local, state, and federal agencies, and Army forces cooperate and synchronize efforts closely with them. These agencies are trained, resourced, and equipped more extensively than similar agencies involved in the conduct of stability tasks overseas. Policies issued by the federal government govern the essential services that Army forces provide in response to disasters. Within this context, a focus for engineers during DSCA will be the restoration of essential services. Combat and general engineering capabilities may be applied to restore essential services. Engineer equipment is well suited for the removal of rubble and debris associated with rescue and access to affected areas. Other likely requirements include the construction of temporary shelters and the provision of water and sanitation services. Likely engineer missions are similar to those required during the conduct of stability tasks, except that they are not conducted outside of U.S. territorial jurisdiction.

Engineer Support to Unified Land Operations

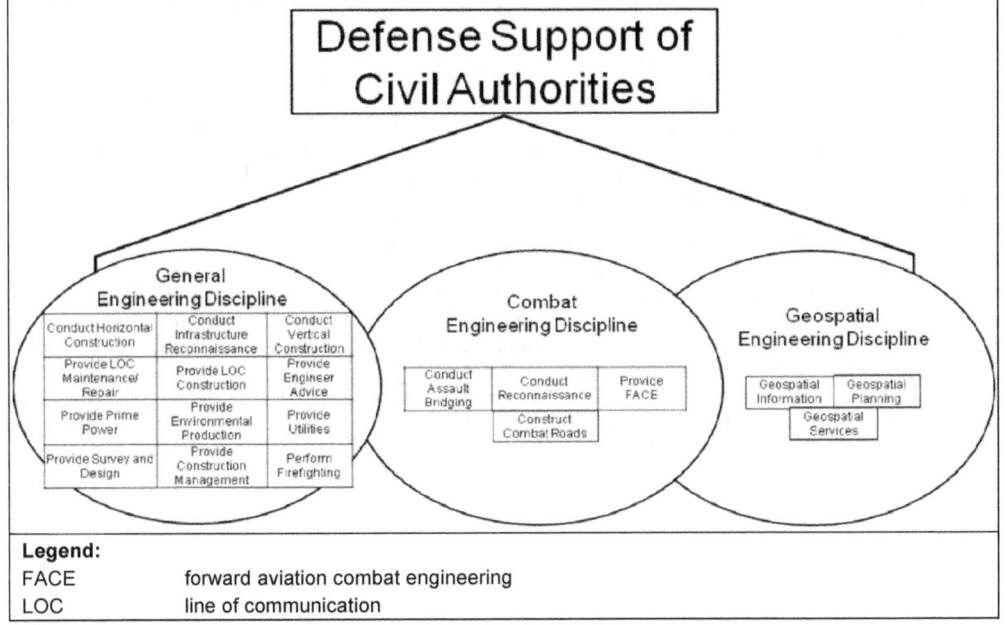

Figure 2-5. Notional engineer support to DSCA tasks

2-106. Engineer support to DSCA may include the typical integration with, and support for, combined arms forces during missions. Combat engineer route clearance and other capabilities may be critical tasks that are applied through the movement and maneuver warfighting function. Geospatial engineering support continues to provide foundational information that supports the common operational picture. General engineering support may be required for the sustainment and protection requirements of the force and may be extended to support other agencies. This may include the following missions:

- Base camp construction and power generation.
- Debris or route clearing operations.
- Construction and repair of roads.
- Forward aviation combat engineering, to include the repair of paved, asphalt, and concrete runways and airfields.
- Installation of assets that prevent foreign object damage to rotary-wing aircraft.
- Construction of temporary bridging.
- Construction and upgrade of ports; airfields; and reception, staging, forward movement, and integration facilities to ensure access to the region.

2-107. DSCA may require an immediate response. USACE maintains significant response capability, and they will normally be involved in providing engineer support to civil authorities. USACE leverages capabilities and expertise developed through responsibility for military construction and civil works programs to prepare for assigned and anticipated DSCA missions.

SPECIAL CONSIDERATIONS

2-108. Army commanders will likely determine that operations in urban environments will be essential to mission accomplishment. They need to assess the relevance and impact of one or more urban areas as part of the mission. They will also need to determine whether urban operations may be the sole focus of the commander or only one of several tasks nested in an even larger operation. Although urban operations potentially can be conducted as a single battle, engagement, or operation, they will more often be conducted as a major operation requiring joint resources. FM 3-06 provides a framework—assess, shape, dominate, and transition—for urban operations. These are not phases or sequential operations, but rather a means to visualize the fight (or potentially, the stability or DSCA mission).

2-109. Geospatial engineers provide a unique form of collaboration with the intelligence staff in the form of geospatial intelligence. Geospatial engineers support geospatial intelligence by providing data standards, data processing, geospatial engineers, geospatial information technicians, and equipment located within a geospatial intelligence cell that can generate, manage, analyze, and disseminate geospatial information and services, enabling the commander to understand the physical environment. (See TC 2-22.7 for a further discussion of the geospatial intelligence cells.)

2-110. Assured mobility will be an important framework for commanders to use as maneuver commanders think about how to shape and dominate within the urban terrain. General engineering tasks will be prevalent throughout operations, but will also be the major function during transition to stability or DSCA tasks. Combat engineers will have to work closely with the elements that enable M/CM/S. They must ensure close coordination with EOD in the reduction of explosive hazards (improvised explosive devices and unexploded ordnance) to minimize collateral damage. Engineers may have to coordinate with military police to enable the movement of civilians along routes or with chemical, biological, radiological, and nuclear elements to detect and identify potential chemical, biological, radiological, and nuclear threats and hazards along routes and other locations within the area of operations.

Chapter 3
Integrating Engineer Support

Engineer planners and staff members in combined arms or other nonengineer headquarters must understand, and become integral members during, the operations process activities at that headquarters. This chapter discusses the operations process as the context for the integration of engineer support to operations. It enters the operations process by discussing various planning activities required for effective engineer support. It describes planning responsibilities, integration, and processes for engineer units and for engineer planners in nonengineer units. Finally, it discusses the preparation, execution, and continuous assessment of the entire spectrum of engineer support.

INTEGRATED PLANNING

3-1. Commanders integrate input from subordinate commanders into the planning process. Engineer leaders must understand, and be integral participants in, the planning process that is affecting the engineer activities at the echelon of employment. Supporting engineer unit commanders and leaders conduct parallel planning that provides effective outcomes for the engineer units that are employed and the appropriate input to the process of the higher headquarters. Geospatial elements and engineer staff planners integrate directly within the planning staff at each echelon to participate in the planning process.

ENGINEER SUPPORT TO THE PLANNING PROCESS

3-2. As a significant part of the planning process, the staff recommends the appropriate command and support relationship between engineer and maneuver units to the commander. Each situation is unique and requires its own solution. Whatever the selected relationship, engineer commanders are inherently responsible for ensuring that engineer support tasks are accomplished by subordinate units. In a command relationship, command authority over engineer units is given to a maneuver commander for the immediate availability of engineer forces when needed. This relationship is well suited for offensive operations and fluid situations, allowing the maneuver commander more flexibility in using engineer assets. Command, administrative, and logistical responsibilities remain with the parent engineer unit in a support relationship. Commanders are assigned a support relationship during the conduct of offense, defense, and stability or DSCA tasks when subordination of one unit to another is inappropriate. The engineer unit commander organizes the unit and suballocates tasks so that they will effectively meet the needs of the maneuver commander.

3-3. Engineer support is complex and resource-intensive; it requires time, manpower, equipment, materials, and extensive and proactive coordination. Additionally, a successful engineering effort requires an understanding of the engineer requirements (combat, general, and geospatial) and the concept of operations roles. Engineer support must be directed and synchronized through planning as one of the critical activities in the operations process, but many engineer activities also require the critical reasoning skills and problem-solving techniques that form the fundamental logic for the planning processes. (See ADP 5-0 for additional information.) Engineer support will involve the use of some functionally unique analytic tools to solve construction, design, facility, and other engineer-specific problems.

3-4. Engineers conduct planning at the strategic, operational, and tactical levels. It is important to understand planning within the context of the levels of war. The scope, complexity, and length of planning horizons differ between operational and tactical planning, yet as echelons of responsibilities have blurred, essentially any engineer headquarters may find itself supporting a maneuver unit at any level of war. For example, an engineer battalion may deploy to support a JTF or an Army corps at the operational level or a division or BCT at the tactical level.

3-5. The engineer planning concepts of the CCDR or senior Army commander focuses on the relationship of geography and force projection infrastructure to the concept of operations. Engineer planners must determine the basic, yet broad, mobilization, deployment, employment, and sustainment requirements of the CCDR concept of operations for the impact on engineer requirements. The senior engineer commander or the engineer staff officer at each echelon must support the development of the supported commander's operation plan (OPLAN) or OPORD and an internal OPLAN or OPORD for the engineer organization. As previously discussed, the engineer staff officer is the special staff officer responsible for coordinating engineer assets and operations for the command, including engineer planning. The engineer staff officer is usually the senior engineer officer on the staff, but it may be a senior engineer commander supporting the force.

3-6. In planning at every level, the engineer planner should consider a number of the following general considerations:

- **Speed.** Engineer tasks are resource-intensive in terms of time, materials, manpower, and equipment. Practices that support speed include the use of existing facilities, standardization, simplicity of design and construction, modular systems, prefabricated or preengineered components, and phased construction.
- **Economy.** Engineering demands the efficient use of personnel, equipment, and materials. Practices that support the economy include the conservation of manpower, equipment, and materials and the application of environmental considerations early in the process.
- **Flexibility.** Standard plans that allow for adjustment, expansion, and contraction will be used when possible. For example, forward airfields should be designed and located so that they can be expanded into more robust facilities.
- **Decentralization of authority.** The dispersion of forces requires that engineer authority be decentralized as much as possible. The engineer commander at a particular location must have authority that is consistent with responsibilities.
- **Establishment of priorities.** Priorities and resource allocation must be established to determine how much engineer effort is devoted to a single task. All levels of command, beginning with the joint force commander, will issue directives establishing broad priorities. Resources are initially assigned to the highest priority tasks, and low priority tasks are left undone while recognizing and mitigating the risk.

ENGINEER ACTIVITIES SPANNING THE LEVELS OF WAR

3-7. The challenges of planning, preparing, executing, and continuously assessing operations within diverse theaters are numerous and varied. The engineer staff must be involved in the operations planning process at each level of war (strategic, operational, and tactical). (See ADP 3-0 for additional information.) Understanding the challenges and opportunities identified from an engineer view equips the staff with relevant information to form a more comprehensive understanding. The omission of engineer considerations at any level may adversely affect the effectiveness of the operation. Engineer support to operations must be synchronized from the strategic level to the tactical level.

Strategic

3-8. Engineer planners must determine the means, ways, and ends as part of a joint force to prevent, shape, and win decisively. Activities include planning the right engineer force with the right mixture of capabilities, engineer policy, and doctrine development in place to mobilize, deploy, employ, sustain, and redeploy forces. Engineer activities at the strategic level seek ways to contribute to preventing, shaping, and winning by setting conditions for decisive action. Engineers conduct force planning, develop engineer policy, and support campaigns and operations, focusing primarily on the means and capabilities to generate, deploy, employ, sustain, and recover forces.

3-9. Additionally, infrastructure development is a critical aspect of enabling and sustaining force deployments, and it places a heavy demand on engineer requirements. Engineers at the strategic level advise on terrain and infrastructure, to include—

- Geospatial information and services.
- Seaports of debarkation.
- Aerial ports of debarkation.
- Force generation.
- Engineer support priorities.
- LOCs.
- Air base and airfield operations.
- Basing strategy.
- Joint targeting.
- Foreign humanitarian assistance.
- Environmental considerations.
- Engineer interoperability.
- Input to the rules of engagement.
- Rules for the use of force.
- Support to protection.

3-10. Environmental issues can have strategic implications. They can also affect mission success and end states if the issues are not recognized early and incorporated into planning and operations. Environmental considerations may include input to the rules of engagement for targeting cultural sites, developing guidance for targeting industrial infrastructure, deciding which laws and treaties pertain to the environmental situation, and determining the level at which the military may conduct environmental remediation and restoration. Natural resource protection can be a key strategic mission objective and is important to HN reconstruction. Failure to recognize environmental hazards can result in significant risk to the JTF, adversely affecting readiness. If not appropriately addressed, environmental issues have the potential to negatively affect local community relations, affect insurgent activities, and create diplomatic problems for the JTF. (See FM 3-34.5 for environmental considerations.)

Operational

3-11. To shape and win decisively, engineer planners must anticipate the impact of geography, force projection infrastructure with specific engineer missions, and available engineer forces within the supported geographic combatant command area of responsibility. Engineer activities at the operational level focus on the impact of geography and force projection infrastructure on the CCDR operational design. Engineer planners must determine the basic, yet broad, mobilization, deployment, employment, and sustainment requirements of the CCDR concept of operations. Engineer planners must secure funding within authorities and plan for procurement of Class IV supplies and services. Operational planning merges the OPLAN or OPORD of the joint force, specific engineer missions assigned, and available engineer forces to achieve success. Combatant command engineer planners also need to understand the capabilities and limitations of Service engineer forces.

3-12. Many of the engineer activities conducted for strategic operations are also performed at the operational level. At the operational level, engineers prioritize limited assets and mitigate risks. Engineers conduct operational area and environmental assessments and work with intelligence officers to analyze the threat. Engineers anticipate requirements and request capabilities to meet them. They provide the scheme of base camps, geospatial products and services, and recommendations on joint fires and survivability for the forces employed. (See ATP 3-37.10 for additional information on base camps.) As the link to tactical engineer integration, operational planners set the conditions for success at the tactical level by anticipating requirements and ensuring that capabilities are available to accomplish engineer support requirements. An example of this includes field forces that are assigned to the operational Army (FESTs, the 249th Engineer Battalion (Prime Power), USACE).

3-13. Engineer staff officers assigned to special operations forces or the Special Forces Command are responsible for planning, coordinating, and executing engineer support. Engineers at this echelon provide policy and direction in the aspects of engineering, to include coordination for engineer support from conventional forces. Due to the nature, scope, and remote environments in which special operations forces operate, theater infrastructure is not always available. Despite recent increases in special operations force structure, conventional force engineers across the three disciplines can be requested to provide additional engineer support. Requests for conventional engineers at this level could be to support special operations in core activities—ranging from augmenting special operation forces in training exercises to providing technical capabilities to restore essential services, to providing infrastructure reconstruction and humanitarian relief, to showing U.S. commitment in the area of interest. Engineers must be familiar with fiscal policy, and they must have the ability to translate special operations requirements in terms that the supporting conventional forces can understand and execute.

Tactical

3-14. Engineer planners must determine the best methods to task-organize forces at the lowest level to support the maneuver of combat forces to win decisively. Engineer activities at the tactical level focus on support to the ordered arrangement and maneuver of forces—in relationship to each other and to the enemy—that are required to achieve combat objectives. At the same time, engineer support is critical to achieving necessary stability tasks.

3-15. Tactical planning in the context of engineer support to operations translates to a primary focus on combat engineering tasks and planning done within tactical organizations. The senior engineer staff officer in the brigade is the primary planner at the tactical level. Engineer tactical planning is typically focused on maneuver support and sustainment support that is not addressed by the higher-echelon commander. Construction planning at the tactical level will typically focus on survivability tasks in support of the protection warfighting function and infrastructure development in support of the sustainment warfighting functions primarily. Engineer planners at the tactical level use the engineer assets provided by operational planners to support the tactical mission tasks assigned to the combat maneuver units they support. With the support of engineers, subordinate commanders ensure that engineering capabilities are effectively integrated into the scheme of maneuver and the performance of assigned tasks. Tactical missions are complex, and planning must consider threat capabilities.

3-16. Special consideration includes performing terrain analysis with an understanding of threat capabilities. Geospatial engineers provide unique graphical representations and terrain data that enable commanders to visualize the area of operations. Engineer reconnaissance (tactical and technical) is a critical capability to the maneuver commander at the tactical level. Threat information must be very specific. Engineers discern and identify patterns and plan specific detection strategies based on the threat. The proliferation of explosive hazards requires engineers to develop new countering procedures continuously. The tactical integration of EOD capabilities has become an increasing requirement. (See ATTP 3-34.80 and FM 3-34.170 for additional information.)

3-17. Engineer support to special operation forces at this echelon has been allocated to provide engineer expertise across the engineer disciplines. Engineer units at this echelon should be prepared to provide an engineer liaison officer to be integrated into the receiving special operations forces headquarters. Engineer planners should be able to provide engineer support that is no different than the support provided to other organizations with the exception that contingency and crisis action planning are the two primary methodologies used. At this level, planning and execution are decentralized. Engineer staff officers must plan for the right personnel and equipment package to conduct engineer operations in austere environments without extensive support until follow-on conventional forces arrive. Engineer organizations do not execute engineer tasks differently than they would for any other decisive action, but they do execute engineer tasks with an emphasis on speed and resource ingenuity.

STAFF PROCESSES

3-18. The Army planning methodologies assist commanders and staffs with effective planning processes. The Army design methodology, military decisionmaking process, and troop leading procedures are three planning processes defined in ADRP 5-0. Leaders determine the appropriate mix based on the mission or operation. Each is a means to an end, and its value lies in the result, not the process. Processes can be performed in detail if time permits or in an abbreviated fashion in a time-constrained environment.

3-19. Although not fully developed planning methodologies, engineers use a number of other processes, activities, and frameworks to facilitate the planning and integration of engineer support. They include—
- The running estimate.
- The framework of assured mobility.
- The development of essential tasks for M/CM/S.

PLANNING

3-20. Except in the smallest echelon of Army units, commanders will rely on assistance from a staff to conduct the planning processes that lead to the OPLAN or OPORD. ADP 6-0 describes the organization and responsibilities of the engineer staff. Engineer planners provide for the integration of engineer-focused considerations on the supported staff at each echelon. Throughout the planning process, the engineer staff must advise supported commanders and staffs about engineering capabilities, methods of employment, and the additional capabilities and depth of the Engineer Regiment. In those units without organic engineer staff support, including support type organizations, it may be important for the supporting engineer organization to provide planning support. Liaison may need to be provided in certain situations to ensure that proper and complete staff planning is accomplished.

3-21. The engineer staff officer at each echelon is responsible for engineer logistics estimates, and the engineer staff officer plans and monitors engineer-related sustainment support for engineering capabilities operating at that echelon. When an engineer unit or capability is task-organized in support of the unit, the engineer staff officer recommends the most effective command or support relationship, including considering the impact of inherent sustainment responsibilities. The engineer staff officer—
- Determines engineer intelligence requirements for an area of operations.
- Writes the engineer annex and associated appendixes to the OPLAN or OPORD to support the commander's intent, to include a recommended distribution for engineer-related, command-regulated classes of supply and special equipment.
- Assists in planning the location of forward supply points for the delivery of engineer-configured loads of Class IV and Class V supplies. This site is coordinated with the unit responsible for the terrain and the appropriate logistics staff officer (S-4) or assistant chief of staff, logistics (G-4).
- Assists in planning the location of the engineer equipment parks for the pre-positioning of critical equipment sets (tactical bridging). This site is coordinated with the unit responsible for the terrain and the appropriate S-4 or G-4.
- Works closely with the sustainment staff to identify available haul assets (including HN) and recommends priorities to the sustainment planners.
- Identifies extraordinary medical evacuation requirements or coverage issues for engineer units and coordinates with sustainment planners to ensure that the supporting unit can accomplish these special workloads.
- Identifies critical engineer equipment and engineer mission logistics shortages.
- Provides the appropriate S-4 or G-4 with an initial estimate of required Class III supplies in support of construction.
- Provides the appropriate S-4 or G-4 with an initial estimate of required Class IV and Class V supplies for the countermobility and survivability efforts.
- Provides the appropriate S-4 or G-4 with an initial estimate of required Class IV supplies in support of construction. Monitors and advises, as required, implications of statutory, regulatory, and command policies for the procurement of construction materials. The critical issue for the engineer staff officer is the timely delivery at required specifications, whatever the source of construction materials.
- Tracks the flow of mission-critical Class IV and Class V supplies into the support areas and forwards the supplies to the supporting engineer units.
- Coordinates engineer assistance, as required, to accept the delivery of construction materials.

Chapter 3

- Coordinates main supply route clearing operations and tracks the status at the main command post.
- Coordinates for EOD support and integration as necessary.
- Serves as the primary staff integrator for the environmental program.

3-22. The staff assigned to the BCT and above includes many engineers in various sections and cells. One of these engineers, typically the senior engineer officer on staff, is designated as the engineer staff officer to advise the commander and assist in exercising control over engineer forces in the area of operations. The engineer staff officer is responsible for coordinating engineer assets and operations for the command. Although there may be more than one engineer officer on a staff, only one is designated as the engineer staff officer for the command. Each echelon, down to the BCT level, has an organic engineer planner and staff element to integrate engineers into the combined arms fight. The task force and company levels may have a designated engineer planner, but the engineer is not typically organic at these echelons. The engineer is a special staff member who is responsible for understanding the full array of engineering capabilities (combat, general, and geospatial) available to the force and for synchronizing them to best meet the needs of the maneuver commander.

3-23. The senior engineer should not be assigned duties as commander and staff officer; however, if serving as the unit commander and engineer staff officer is considered, some specific considerations for determining the relationship of the senior engineer staff officer and the engineer unit commander should include—

- What staff assets are available to support the engineer staff advisor versus the engineer unit commander? Are the elements from the same unit, or are separate units resourced for each role?
- What experience level is needed for the engineer staff advisor? Should this role be resourced with a current or former commander?
- What duration of time will the augmenting engineer element, commanded by the senior engineer unit commander, be working for or with the force? Does the engineer commander have the time to acclimate and effectively advise the force commander?
- What working relationship is established between an existing engineer staff advisor and the force commander? Similarly, is there an existing working relationship between the engineer unit commander and this force commander? It is critical that the engineer staff officer for the supported unit maintains close coordination with the supporting engineer unit commander and staff to ensure a synchronization of effort.

3-24. The engineer staff will include key members on many of the working groups, boards, or cells established by commanders to coordinate functional or multifunctional activities. The engineer staff officer may chair construction-related groups.

3-25. The specific roles, responsibilities, and considerations for the engineer staff officer are similar, but not identical, at each echelon. FM 3-34.22 addresses these for the BCT engineer staff officer, while ATTP 3-34.23 addresses them for the engineer staff officer at echelons above the BCT.

3-26. Successful sustainment of engineer organizations and capabilities requires active involvement by engineer commanders and staffs at every echelon. In addition to ensuring the sustainment of the units, engineers must work closely with supported units. This is because the supported unit is responsible for providing the Class IV and Class V construction and obstacle material needed for the tasks they assign to the supporting engineer unit, regardless of the command and support relationship between them. The higher-echelon engineer staff officer must retain an interest in the sustainment of subordinate engineer units and capabilities, regardless of the command and support relationships with the supported units. Within a supported unit, the engineer staff officer must work closely with the logistics staff to assist in planning, preparing, executing, and assessing operations requiring engineer materials and resources. Within engineer or multifunctional headquarters units, the logistics staff provides sustainment planning for its subordinate units.

3-27. Within engineer units, leaders and staff must monitor, report, and request requirements through the correct channels and ensure that sustainment requirements are met when sustainment is brought forward to the engineer unit. The accurate and timely submission of personnel and logistics reports and other necessary information and requests is essential.

3-28. Engineer commanders and the engineer staff officer must ensure that parallel planning occurs between the supported unit and the task-organized engineer units. This parallel process feeds into the force commander's military decisionmaking process and provides input for an engineer unit OPLAN, OPORD, or annex to be published nearly simultaneously, maximizing the time available for execution.

3-29. To facilitate effective parallel planning at the engineer unit level, engineer unit commanders and staff planners must—
- Understand the commander's intent and planning guidance of the parent (engineer) unit and the supported unit.
- Analyze the terrain, obstacle information, and threat capabilities.
- Know the engineer systems and capabilities to accomplish the identified tasks within the time allotted.
- Identify risks where engineering capabilities are limited or time is short, and identify methods to mitigate the risks ensuring that potential reachback capabilities have been leveraged.
- Consider the depth of the area of operations and the transitions that will occur among operational elements. This includes the integration of environmental considerations.
- Plan for the sustainment of engineer activities. Engineers ensure that the logistical requirements are analyzed and accounted for through the end state and resourced to accomplish the mission and facilitate future operations.

MILITARY DECISIONMAKING PROCESS

3-30. Engineers analyze the operational environment using operational variables to add to the shared common understanding by identifying potential challenges and opportunities within the operation before and during mission execution. The resulting understanding of the operational environment (an engineer view of the operational environment) is not intended to be limited to considerations within the operational environment that may result in engineer functional missions. The resulting engineer view of the operational environment is, instead, organized by lines of engineer support and linked to the common overall understanding through the warfighting functions.

Operational Variables

3-31. Army doctrine describes an operational environment in terms of the following eight constantly interacting operational variables—political, military, economic, social, information, infrastructure, physical environment, and time (PMESII-PT). The following examples are provided to show the added focus sought within each of the operational variables by the engineer view of the operational environment. The examples are not meant to restate the more complete treatment of the variable in the general terms provided in ADRP 5-0 or to be an all-inclusive treatment of the engineer aspects within each of the variables, but to focus engineer perspectives on—
- **Political.** Understanding the political circumstances within an operational environment will help the commander recognize key actors and visualize explicit and implicit aims and capabilities to achieve goals. The engineer view might add challenges associated with political circumstances that permit or deny access to key ports of entry or critical sustainment facilities. Opportunities in the form of alternative access routes might be added. The engineer and others may be impacted by the effect of laws, agreements, or positions of multinational partners (restrictions on shipment of hazardous material across borders or a host of similar political considerations that can affect engineer planning and operations).
- **Military.** The military variable explores the military capabilities of relevant actors in a given operational environment. The engineer view might add the challenges associated with an enemy capability to employ explosive hazards or other obstacles and the capability to challenge traditional survivability standards. Opportunities in the form of existing military installations and other infrastructure would be added. The engineer view includes a necessarily robust and growing understanding of engineering capabilities in a context of unified action within this variable of the operational environment.

- **Economic.** The economic variable encompasses individual behaviors and aggregate phenomena related to the production, distribution, and consumption of resources. The engineer view might add challenges associated with the production or availability of key materials and resources. Opportunities in the form of potential for new or improved production facilities might be added.
- **Social.** The social variable describes the cultural, religious, ethnic makeup, and social cleavages within an operational environment. The engineer view might add challenges associated with specific cultural or religious buildings or installations, the impact of language barriers or availability of laborers, and qualified local engineer resources. Opportunities in the form of potential to provide for culturally related building requirements might be a consideration.
- **Information.** This variable describes the nature, scope, characteristics, and effects of individuals, organizations, and systems that collect, process, disseminate, or act on information. Engineers assist the commander by informing and influencing activities to shape the operational environment through the capability to improve infrastructure and services for the population. The engineer must consider how construction projects, especially in stability operations, will support informational themes that are consistent with friendly military goals and actions. The engineer view might also add challenges associated with deficiencies in the supporting architecture, to include power considerations.
- **Infrastructure.** Infrastructure comprises the basic facilities, services, and installations needed for the functioning of a community or society. The engineer view might add challenges associated with specific deficiencies in the basic infrastructure. Opportunities in the form of access to existing infrastructure, improvements to existing infrastructure, and new projects might be added. The engineer view provides for a detailed understanding of infrastructure by using sewage, water, electricity, academics, trash medical, safety, and other considerations (SWEAT-MSO). See FM 3-34.170 and FM 3-34.400 for additional information.
- **Physical environment.** The defining factors are urban settings (supersurface, surface, and subsurface features) and other types of complex terrain, weather, topography, hydrology, and environmental conditions. An enemy may try to counteract U.S. military advantages by operating in urban or other complex terrain requiring greater engineer effort to provide the freedom of action. The engineer view might add challenges associated with natural and man-made obstacles. Insights into environmental considerations are also a concern. (See FM 3-34.5 for additional information.) Opportunities in the form of existing routes, installations, and resources might be added. The engineer view supports a broad understanding of the physical environment through geospatial engineering, which is discussed in detail in ATTP 3-34.80 and JP 2-03.
- **Time.** The variable of time influences military operations within an operational environment in terms of the decision cycles, operational tempo, and planning horizons. The duration of an operation may influence engineer operations in terms of whether to pursue permanent or nonpermanent base camp solutions for facilities and infrastructure. The CCDRs establish base camp strategy that is tailored to the joint operational area based on an assessment of the situation, unique characteristics of the region, and anticipated duration.

Mission Variables

3-32. While an analysis of the operational environment using the operational variables (PMESII-PT) improves situational understanding, when commanders receive a mission, they require a mission analysis focused on the specific situation. The Army uses the mission variables as the categories of relevant information used for mission analysis. Similar to the analysis of the operational environment using the operational variables, the engineer uses the mission variables to seek the shared common understanding from an engineer perspective.

3-33. The following are some examples of the engineer perspective for each of the mission variables:
- **Mission.** Commanders analyze a mission in terms of specified tasks, implied tasks, and the commander's intent (two echelons up) to determine essential tasks. Engineers conduct the same analysis, with added focus on the engineer requirements, to determine the essential tasks and engineer priorities. The early identification of the essential tasks for engineer support enables the maneuver commander to request engineer augmentation early on in the planning process.
- **Enemy.** The engineer view of the enemy concentrates on enemy tactics, equipment, and capabilities that could threaten friendly operations. This may include an analysis of enemy disposition, enemy engineering capabilities, obstacle intelligence, engineer reconnaissance, and mine strike reporting within the area of operations or area of interest that could have an impact on mission.
- **Terrain and weather.** As the terrain visualization experts, geospatial engineers analyze terrain (man-made and natural) to determine the effects on friendly and enemy operations. Geospatial engineers analyze terrain using the five military aspects of terrain (observation and fields of fire, avenues of approach, key terrain, obstacles, and cover and concealment [OAKOC]). Geospatial engineers integrate geospatial products to help commanders and staffs visualize the terrain.
- **Troops and support available.** Engineers consider the number, type, capabilities, and condition of available engineer troops and support available from unified action partners.
- **Time available.** Engineers must understand the time required in planning engineer operations and the importance of collaborative and parallel planning to prepare and execute tasks. Engineers realize the time needed for positioning critical assets and the time associated with performing engineer tasks or projects.
- **Civil considerations.** The influence of man-made infrastructure; civilian institutions; and attitudes and activities of the civilian leaders, populations, and organizations within the area of operations impact the conduct of military operations. At the tactical level, they directly relate to key civilian areas, structures, capabilities, organizations, people, and events (ASCOPE). This engineer view provides a detailed understanding of the basic infrastructure needed for a community or society. The engineer view might identify challenges, to include environmental stewardship, financial and economic feasibility, social and cultural impacts, and the implications associated with specific deficiencies in the basic infrastructure and opportunities for improvement or development of the infrastructure.

Engineer Staff Running Estimate

3-34. The engineer staff officer uses the running estimate as a logical thought process and as an extension of the military decisionmaking process. It is conducted by the engineer staff officer, concurrently with the planning process of the supported force commander, and is continually refined. This estimate allows for the early integration and synchronization of engineer considerations into combined arms planning processes. In running estimates, staff sections continuously consider the effect of new information and update assumptions, friendly force status, effects of enemy activity, civil considerations, and conclusions and recommendations. A section running estimate assesses the following:
- Friendly force capabilities with respect to ongoing and planned operations.
- Enemy capabilities as they affect the section area of expertise for current operations and future plans.
- Civil considerations as they affect the section area of expertise for current operations and future plans.
- The operational environments effect on current and future operations from the section perspective.

3-35. The development and continuous maintenance of the running estimate drive the coordination between the staff engineer, supporting engineers, the supported commander, and other staff officers in the development of plans, orders, and the supporting annexes. Additionally, the allocation of engineer assets and resources assists in determining command and support relationships that will be used. Table 3-1, page 3-10, shows the relationship between the military decisionmaking process and the engineer staff running estimate.

Table 3-1. Military decisionmaking process and the engineer staff running estimate

Military Decisionmaking Process	Engineer Staff Running Estimate

Mission analysis:
• Analyze the higher headquarters plan or order.
• Perform the initial IPB.
• Determine the specified, implied, and essential tasks.
• Review the available assets, and identify resource shortfalls.
• Determine the constraints.
• Identify the critical facts, and develop assumptions.
• Begin the composite risk assessment.
• Determine the CCIR and EEFI.
• Develop the information collection plan.
• Update the plan for the use of available time.
• Develop the initial information themes and messages.
• Develop the proposed mission statement.
• Present the mission analysis briefing.
• Develop and issue the initial commander's intent.
• Develop and issue the initial planning guidance.
• Develop the COA evaluation criteria.
• Issue the warning order.
	Analyze the mission:
	• Analyze the higher headquarters orders.
	  ▪ Commander's intent.
	  ▪ Mission.
	  ▪ Concept of operation.
	  ▪ Timeline.
	  ▪ Area of operations.
	• Conduct the IPB, and develop the engineer staff running estimate.
	  ▪ Terrain and weather analysis.
	  ▪ Enemy mission and M/CM/S capabilities.
	  ▪ Friendly mission and M/CM/S capabilities.
	• Analyze the engineer mission.
	  ▪ Specified M/CM/S tasks.
	  ▪ Implied M/CM/S tasks.
	  ▪ Available assets.
	  ▪ Limitations.
	  ▪ Risk as applied to engineering capabilities.
	  ▪ Time analysis.
	  ▪ Essential tasks for M/CM/S.
	  ▪ Restated mission.
	• Conduct the risk assessment.
	  ▪ Safety.
	  ▪ Environment (EBS/OEHSA).
	• Determine the terrain and mobility restraints, obstacle intelligence, threat engineering capabilities, and critical infrastructure.
	• Recommend the CCIR.
	• Integrate the engineer reconnaissance effort.

COA development	Develop the scheme of engineer operations. • Analyze the relative combat power. • Refine the essential tasks for M/CM/S. • Identify the engineer missions and the allocation of forces and assets. • Determine the engineer priority of effort and support. • Refine the commander's guidance for M/CM/S operations. • Apply the engineer employment considerations. • Integrate engineer support into the maneuver COA. (See ATTP 5-0.1 for additional information on the scheme of engineer operations.)

Table 3-1. Military decisionmaking process and the engineer staff running estimate (continued)

Military Decisionmaking Process	Engineer Staff Running Estimate
COA analysis	War-game and refine the engineer plan.
COA comparison	Recommend a COA.
COA approval	Finalize the engineer plan.
Orders production, dissemination, and transition	Create the input to the basic operation order. • Scheme of engineer operations. • Essential tasks for M/CM/S. • Subunit instructions. • Coordinating instructions. Create the engineer annex and appendixes.
Legend: CCIR commander's critical information requirements COA course of action EBS environmental baseline survey EEFI essential elements of friendly information IPB intelligence preparation of the battlefield M/CM/S mobility, countermobility, and survivability OEHSA occupational environmental health site assessment	

PLANS AND ORDERS

3-36. The staff prepares the order or plan by turning the selected course of action into a clear, concise concept of operations with the required supporting information. The concept of operations for the approved course of action becomes the concept of operations for the plan. The course-of-action sketch becomes the basis for the operation overlay. Orders and plans provide information that subordinates need for execution. Mission orders avoid unnecessary constraints that inhibit subordinate initiative. The staff assists subordinate unit staffs with planning and coordination.

3-37. The engineer staff planner provides input for the appropriate paragraphs in the base plan and the annexes and appendixes of the base plan as found in ATTP 5-0.1. In addition to developing input for the functionally specific paragraphs, engineer planners must review other sections as well. Engineers ensure the integration of geospatial support in the appropriate sections and annexes. Engineers review the task organization to ensure sufficient capability to meet identified requirements. The engineer planner recommends the appropriate command or support relationships. Additionally, planners provide input to the flow of the engineer force as detailed on the time-phased force and deployment data. Engineers review operations sections, annexes, and overlays to ensure the inclusion of obstacle effects or other graphics and assist in conveying the scheme of engineer operations. In the fires section, engineers work with the fire support officer and other members of the staff to integrate obstacles with fire. Employing scatterable mines and confirming that obstacles are covered by fire are of particular interest.

3-38. An engineer annex, normally found in annex G of the base plan or base order is the principal means through which the engineer defines engineer operations to the maneuver commander's intent, essential tasks for M/CM/S, and coordinating instructions to subordinate commanders. It is not intended to function as the internal order for an engineer organization, where the engineer commander articulates intent, the concept of operations, and coordinating instructions to subordinate, supporting, and supported commanders. The preparation of the annex seeks to clarify the scheme of engineer operations to the OPLAN or OPORD and includes the—

- Overall description of the engineer staff officer scheme of engineer operations, including approved essential tasks for M/CM/S.
- Priorities of work to shape the theater or area of operations (not in a tactical level engineer annex).

Chapter 3

- Operational project planning, preparation, and execution responsibilities (not in a tactical level engineer annex).
- Engineer organization for combat.
- Essential tasks for M/CM/S for subordinate units.
- Allocations of Class IV and Class V (obstacle material).

Note. Guidance to maneuver units on obstacle responsibilities should be listed in the body of the basic order, not in the engineer annex.

3-39. The engineer staff officer may produce an engineer overlay in conjunction with the operations overlay to highlight obstacle information or breaching operations. A gap-crossing operation may require a separate annex as part of the base order.

3-40. The engineer staff officer performs as the staff integrator and advisor to the commander for environmental considerations. An environmental considerations appendix parallels guidance from the joint OPLAN, OPORD, or concept plan. (See FM 3-34.5 for an example of this appendix.) When specific command procedures dictate, other staff officers include some environmental considerations in logistics and medical annexes. Unit planning at the Regiment or brigade level and below will normally include only those elements required by the higher headquarters orders or plans that are not already included in a unit standing operating procedure. If this appendix is not written, appropriate material will be placed in the coordinating instructions of the basic order.

Sustainment Planning Considerations

3-41. Sustainment support for engineers is provided by different organizations based on various factors, such as the echelon of the supported unit and command and support relationships. Although engineers should be familiar with the sustainment organizations described in ADP 4-0, some organizations provide support to engineers more frequently than others.

3-42. The engineer staff officer, engineer unit commander, supported unit logistics officer, and supporting sustainment unit work closely to synchronize sustainment for engineering capabilities. When the supported unit receives a warning order (directly or implied) as part of the military decisionmaking process, the engineer staff officer initiates the engineer portion of the logistics estimate process. The engineer staff officer focuses the logistics estimate on the requirements for the upcoming mission and the sustainment of subordinate engineer units that are organic and task-organized in support of the unit. Class I, III, IV, and V supplies and personnel losses are the essential elements in the estimate process. Close integration with the sustainment support unit can simplify and accelerate this process using the automated systems logistics status report to ensure that the sustainment support unit is able to maintain an up-to-date picture of the engineer unit sustainment requirements. During continuous operations, the estimate process supporting the rapid decisionmaking and synchronization process may need to be abbreviated due to time constraints.

3-43. The engineer staff officer uses the running estimate to determine the requirements for unit and mission sustainment and compares the requirements with the reported status of subordinate units to determine the specific amount of supplies needed to support the operation. These requirements are then coordinated with the supporting sustainment unit or forward support element to ensure that the needed supplies are identified and resourced. The engineer staff officer then translates the estimate into specific plans that are used to determine the supportability of supported unit courses of action. After a course of action is selected, the specific sustainment input to the supported unit base OPORD and paragraph 4 of the engineer annex is developed and incorporated.

3-44. Engineers use mission type orders and standardized procedures to contribute to simplicity. Engineer commanders and staffs establish priorities and allocate classes of supplies and services to simplify sustainment operations. They use preconfigured loads of specialized classes of supply to simplify transport. At some level, and to some degree, resources are always limited. When prioritizing and allocating resources, the engineer commander may not be able to provide a robust support package. The priority of effort will be established while balancing the mitigation of risk to the operation. Engineer commanders may have to improvise to meet the higher intent and mitigate the risks. Commanders consider economy in prioritizing and allocating resources. Economy reflects the reality of resource shortfalls, while recognizing the inevitable friction and uncertainty of military operations. Engineers must protect the resources needed to sustain units and accomplish missions. In addition to protecting units, personnel, and equipment, engineers must also emphasize security and protection for Class IV and Class V supplies. These supplies are not easily replaced and can be tempting targets for enemy action.

3-45. In each of the different types of BCTs, the engineer staff officer, working with the appropriate sustainment planner and executor, tracks essential sustainment tasks involving engineer units that are supporting the brigade. Accurate and timely status reporting assists the engineer staff officer in providing the overall engineer status to the brigade commander and allows the engineer staff officer to intercede when there are critical sustainment problems. The engineer staff officer also ensures that supplies needed by augmenting EAB engineer units to execute missions for the brigade are integrated into the brigade sustainment plans. For the engineer staff officer to execute these missions properly, accurate and timely reporting and close coordination are essential between the engineer staff officer, sustainment planners and providers, and other engineers within or supporting the BCT. Supporting EAB engineer units must affect linkup with the existing engineer sustainment to ensure the synchronization of effort. Some important considerations for engineer planners include—

- Coordinating for a field maintenance team to support each engineer unit and ensure the quick turnaround of maintenance issues.
- Coordinating closely with the logistics staff to assist in the management of required construction materials. The engineer staff helps the logistics staff identify and forecast requirements to ensure that a quality control process is in place for the receipt of the materials. The management of Class IV supplies for survivability and countermobility is most efficient when there is a shared interest between the maneuver and engineer logisticians.
- Using expeditionary support packages of barrier materials.
- Coordinating closely with the theater support command or sustainment command support operations officer, the Army forces G-4, the supporting contract support brigade, and the associated logistics civil augmentation program planner to ensure that engineer requirements are properly integrated and captured in the contracting support plan and/or specifically addressed in the engineer support plan.

3-46. Engineers must consider the environmental impacts of their actions. They must ensure that environmental-related risks are included in the risk management process before beginning any action. (See FM 3-34.5 or GTA 05-08-002 for additional information.)

Sustainment Challenges for Engineer Support

3-47. Many sustainment challenges are common to all units, but engineer units face several unique sustainment challenges. Engineers and staffs that employ engineer units and capabilities need to thoroughly understand, anticipate, and work to overcome these challenges.

3-48. Many engineer tasks require the use of engineer equipment that is large and heavy. Low-density haul assets are required to move long distances. Engineer equipment often exceeds size and weight restrictions, making movement more challenging. Much engineer equipment is also low-density, which poses challenges to maintenance and repair. Obtaining engineer-specific Class IX repair parts often requires extraordinary coordination. Mechanics that are capable of maintaining and repairing engineer equipment may also be in short supply, adding to the difficulty in keeping engineer equipment operating.

3-49. Engineer equipment consumes large amounts of fuel (higher than most equipment found in an infantry BCT or Stryker BCT). Refueling is often complicated by the fact that engineer equipment cannot easily travel to refueling points. Time spent traveling between work sites and refueling points can significantly reduce productivity, but bringing fuel trucks to work sites can also be difficult. It is especially difficult when the work sites are scattered over large distances in difficult terrain, and it increases the risk for fuel trucks. It also reduces the availability of fuel trucks for other critical missions.

3-50. Engineer tasks frequently require large amounts of Class IV and Class V supplies. Survivability and countermobility tasks require fortification and barrier materials, mines, and demolitions, while mobility operations require demolitions and construction materials. Construction projects often require significantly large amounts of construction materials. These materials are typically difficult to move, and they require a large commitment of transportation and material-handling equipment support, security, protection, and control. This also places additional demands on other resources.

3-51. Construction materials often require long lead times and can be difficult to acquire in the required quantities and specifications. Statutory, regulatory, and command policies may dictate the source of construction materials, which may require the maximum use of local procurement.

3-52. The frequent movement within an area of operations and the likely changes to task organization and command and support relationships is another challenge. Limited engineer assets often require that they be frequently shifted throughout the area of operations to meet the mission requirements. These movements and changes often have a ripple effect in the sustainment system, which may have difficulty keeping up with these types of multiple changes. This is exacerbated when, as is often the case, engineer missions are conducted in austere environments while infrastructure is being established or improved.

3-53. The requirements for engineer units and assets usually exceed the capacity of available engineer units. This inevitably imposes pressure to delay preventive maintenance checks and services to avoid work stoppages, which increases the likelihood and length of future equipment failures and further compounds maintenance difficulties. This frequently leads to the procurement of locally available construction materials, repair parts, and construction services. This brings unique challenges and the need for financial management and contract management support. Most engineer units do not have dedicated contingency contracting teams, and this support is provided on a general service basis from the supporting contracting support brigade (or joint command if established).

3-54. Some key differences between contracted and military support include—
- Contractor personnel who are authorized to accompany the force in the field are not combatants or noncombatants.
- Contractors are not in the chain of command. They are managed through the contract and the contract management system, which should always include a unit contracting officer representative.
- Contractors perform only the tasks specified in the contract and by the terms of the contract.

3-55. These challenges are predictable and none of them should catch engineer leaders by surprise. Engineers must anticipate these challenges, work to prevent them, and be prepared to overcome them. Because of the critical impact that sustainment has on engineer missions, engineer commanders must be thoroughly familiar with sustainment doctrine and organizations as described in ADP 4-0 and subordinate publications. The importance and unique challenges of contracted support require engineer commanders to fully understand the planning role managing contracted support. (See FM 4-92 and ATP 4-94 for additional information.)

3-56. Engineers must integrate sustainment with engineer plans. Sustainment must not be an afterthought. Engineers must coordinate and synchronize operations with the elements of sustainment. This must occur at all levels of war and throughout the operations process at all echelons. Engineer planners evaluate the sustainment significance of each phase of the operation during the entire planning process. They create a clear and concise concept of support that integrates the commander's intent and concept of operation. This includes analyzing the mission; developing, analyzing, war-gaming, and recommending a COA; and executing the plan. (Table 3-2 lists some engineer planning considerations.)

Table 3-2. Engineer considerations in the military decisionmaking process

MDMP Steps	Engineer Considerations
Receipt of the mission	<ul><li>Receive higher headquarters plans, orders, and construction directives.</li><li>Understand the commander's intent and time constraints.</li><li>Request geospatial information about the area of operations.</li><li>Establish engineer-related boards as appropriate.</li></ul>

Table 3-2. Engineer considerations in the military decisionmaking process (continued)

MDMP Steps	Engineer Considerations
Mission analysis	• Analyze the available information on existing obstacles or limitations. Evaluate terrain, climate, and threat capabilities to determine the potential impact on M/CM/S. • Develop the essential tasks for M/CM/S. • Identify the available information on routes and key facilities. Evaluate LOC, aerial port of debarkation, and seaport of debarkation requirements. • Determine the availability of construction and other engineering materials. • Review the availability of engineering capabilities, to include Army, joint, multinational, HN, and contracted support. • Determine the bed-down requirements for the supported force. Review theater construction standards and base camp master planning documentation. Review unified facilities criteria, as required. • Review the existing geospatial data on potential sites, conduct site reconnaissance (if possible) and environmental baseline surveys (if appropriated), and determine the threat (to include environmental considerations and explosive hazards). • Obtain the necessary geologic, hydrologic, and climatic data. • Determine the level of interagency cooperation required. • Determine the funding sources, as required. • Determine the terrain and mobility restraints, obstacle intelligence, threat engineering capabilities, and critical infrastructure. Recommend the commander's critical information requirements. • Integrate the reconnaissance effort.
COA development	• Identify the priority engineer requirements, including essential tasks for M/CM/S developed during mission analysis. • Integrate engineer support into COA development. • Recommend an appropriate level of protection effort for each COA based on the expected threat. • Produce construction designs that meet the commander's intent. (Use the Theater Construction Management System when the project is of sufficient size and scope.) • Determine alternate construction locations, methods, means, materials, and timelines to give the commander options. • Determine real-property and real estate requirements.
COA analysis	• War-game and refine the engineer plan. • Use the critical path method to determine the length of different COAs and the ability to crash the project.
COA comparison	• Determine the most feasible, acceptable, and suitable methods of completing the engineering effort.
COA approval	• Determine and compare the risks of each engineering COA. • Gain approval of the essential tasks for M/CM/S and construction management, safety, security, logistics, and environmental plans, as required.
Orders production, dissemination, and transition	• Produce construction directives, as required. • Provide input to the appropriate plans and orders. • Ensure that resources are properly allocated. • Coordinate combined arms rehearsals, as appropriate. • Conduct construction prebriefings. • Conduct preinspections and construction meetings. • Synchronize the construction plan with local and adjacent units. • Implement protection construction standards, including requirements for security fencing, lighting, barriers, and guard posts. • Conduct quality assurance and midproject inspections. • Participate in engineer-related boards. • Maintain as built and red line drawings. • Project turnover activities.

Table 3-2. Engineer considerations in the military decisionmaking process (continued)

Legend:	
COA	course of action
HN	host nation
LOC	line of communication
M/CM/S	mobility, countermobility, and survivability
MDMP	military decisionmaking process

CONSIDERATIONS FOR UNIFIED LAND OPERATIONS

3-57. During combat operations, engineer units will tend to have command and support relationships to maneuver commanders. (See ADRP 6-0 for additional information.) Especially at higher echelons, engineer units are more likely to be attached than placed under operational control for a given offensive mission because it lets the gaining unit task-organize and direct engineer forces. Although the forms of offensive maneuver have different intentions, the planning phase must always begin with predicting enemy intent through a thorough understanding of the threat, engineer capabilities, and the effect of terrain on operations. Geospatial products and information become the foundation and common reference for planning. Of all forms of maneuver, the knowledge of enemy disposition is especially critical and required for an infiltration or penetration due to the requirements for stealth and surprise. Engineer planning tends to focus on mobility support, including a robust reconnaissance effort. See FM 3-34.170 for a full discussion of engineer reconnaissance. A greater degree of planning is required for a penetration from the breach to the ultimate control of the objective.

3-58. Planning for defensive operations is inextricably linked to offensive operations and, for planning purposes, consider the transition from offensive operations and the follow-on offensive operations. During the conduct of defensive tasks, engineers use geospatial products to best position units within the defense. Engineers then work with intelligence staffs to describe the threat and to predict where the enemy is likely to attack friendly forces. Engineers also work in conjunction with intelligence staffs to determine which sensor capabilities to leverage and can best predict and prevent the enemy from maneuvering freely into the defended area. Defensive planning includes security and survivability considerations. The consideration of counterattack planning or support for the mobile strike force is the same as the typical mobility planning for offensive operations. The engineer staff officer works with the other staff members to ensure that the counterattack force can mass its effects on the enemy for decisive operations. The type of defensive operation will help to define the amount and focus of engineer effort required. An area defense will typically require a more robust engineer effort due to an increased survivability requirement. A mobile defense effort will typically be less (although mobility requirements may increase) because it has greater flexibility and takes advantage of the terrain in depth.

3-59. The conduct of stability tasks emphasizes nonlethal construction tasks performed by Soldiers who are normally working among noncombatants and local populations. In planning to conduct stability tasks, engineers consider requirements necessary for the support of the primary stability tasks. Engineers are typically critical enablers and may lead in the restoration of essential services. The planner considers capabilities needed to establish or restore the most basic services for the provision of water, emergency shelter, and basic sanitation (sewage and garbage disposal). Terrain products continue to have a great deal of importance, but political and cultural considerations are equally important. Terrain analysts will work with the intelligence staff to develop usable products for the commander to reflect this information where available. When analyzing the troops available, the engineer staff officer considers unified action partner engineering capabilities as a whole, not simply those assigned to the organization. Interaction with these other parties requires engineers to address interoperability, common standards, and mutual agreements. Combined arms forces have a major role in this interaction, working with and through HN agencies and other civilian organizations to enhance the HN government legitimacy.

3-60. Planning DSCA is significantly different from planning offense, defense, or stability tasks because of the unique nature of the hazard or threat, although the basic missions may be very similar to those associated with the conduct of stability tasks. The hazard or threat will likely be a natural or man-made disaster with unpredictable consequences. Additionally, planners must be aware of the number of statutes and regulations that restrict the Army interaction with other government agencies and civilians during DSCA. Local and state responses normally lead the effort, with a federal response providing support as required. Interagency response during DSCA tasks is governed by the *National Response Framework*, which delegates responsibility to various federal agencies for emergency support function. The USACE and other engineering capabilities of the generating force will have a prominent role in DSCA tasks. See ADP 3-28 for additional information about DSCA and the *National Response Framework* emergency support functions.

3-61. As a military partner during DSCA, Army commanders will assume a support role to one or more designated agencies. Engineers can expect to be involved in planning for the support of relief operations with geospatial products and an analysis of potential areas to establish life-support areas. Engineers may be called on to provide manpower support or general engineering support from units with unique capabilities (water well drilling, temporary shelter, power generation, firefighting). Engineer commanders and staff will work with the planners to identify requirements and plan engineer applications. See ATTP 3-34.23 for a more detailed discussion of engineering capabilities used during DSCA.

Engineer Support to Unified Land Operations

3-62. The Army operational concept is unified land operations. *Unified land operations* describes how the Army seizes, retains, and exploits the initiative to gain and maintain a position of relative advantage in sustained land operations through simultaneous offensive, defensive, and stability operations in order to prevent or deter conflict, prevail in war, and create the conditions for favorable conflict resolution. (ADP 3-0).

3-63. Organic engineering capabilities in each of the BCTs provide close support to the maneuver of those forces. Based on a mission variable analysis, the BCTs will be task-organized with additional modular engineering capabilities to meet mission requirements. For offensive and defensive operations, engineer augmentation may consist of additional combat engineering capabilities and an engineer battalion headquarters to provide the necessary mission command for the mix of modular engineer units and capabilities augmenting the BCT. Other, more technically specialized engineering capabilities support the BCT requirements related to the movement and maneuver, protection, and sustainment warfighting functions. These same capabilities may be employed at division, corps, and theater army echelons to enable force mobility, survivability, and sustainment primarily. Force-tailored engineering capabilities from the force pool can provide critical nonlethal capabilities to conduct or support stability or DSCA tasks. Geospatial engineering capabilities, organic and from the force pool, provide support by adding to a clear understanding of the physical environment.

3-64. Engineering capabilities are a significant force multiplier in unified land operations, facilitating the freedom of action necessary to meet mission objectives. Decisive action requires simultaneous combinations of offensive, defensive, and stability or DSCA tasks. Higher-echelon engineer activities are intrinsically simultaneous—supporting combinations of operational components, occurring at every echelon, influencing each level of war, and influencing the entire range of military operations. Engineer activities modify, maintain, provide an understanding of, and protect the physical environment. In doing so, they enable the mobility of friendly forces and alter the mobility of a threat. This enhances survivability; enables the sustainment of friendly forces; contributes to a clear understanding of the physical environment; and provides support to noncombatants, other nations, and civilian authorities and agencies. Indeed, engineer activities may be so widespread and enveloping that they may be viewed as a stand-alone objective, but they are not stand-alone. Engineer applications are effective within the context of the supported objective. Military engineer support is focused on the combined arms objective. To identify and maintain that focus for the widespread application of engineering capabilities, engineer support is integrated within the combined arms operation.

Engineer Support Across the Range of Military Operations

3-65. While violence varies across the range of military operations, the magnitude of requirements for engineers may remain consistently high from peace through war. This demand results in the application of the engineer disciplines to provide a menu of actions available to support military operations.

Chapter 3

3-66. Engineer requirements to support peace during military operations may include geospatial engineering support to provide a clear understanding of the physical environment. Military engagement, security cooperation, and deterrence activities sometimes require large numbers of forces. These forces will need infrastructure, facilities, LOCs, and bases or base camps to support sustainment. Even in areas with well-developed existing infrastructure, significant engineer effort will often be required to plan, design, construct, acquire, operate, maintain, or repair that infrastructure to support operations in-theater. Assistance in response to disaster and humanitarian relief usually includes significant engineering challenges and opportunities to affect the situation immediately and positively.

3-67. Engineer activities to support war during military operations require support for ground combat (or the possibility of ground combat). These require integrating engineer and other support activities with fires and maneuver of ground combat forces to assure the mobility of friendly forces, alter the mobility of threats, and enhance the survivability of friendly forces. It also involves significant challenges associated with sustaining the operation.

3-68. Between peace and war, engineers are often required to improve stability through projects that develop infrastructure and create or improve HN technological capacity. There may also be requirements to provide specialized engineer support to other agencies. Engineers involved in unconventional warfare help overcome challenges to the commander's ability to move and maneuver freely, protect the forces employed, and sustain the operation. Other requirements include directly impacting threat freedom of action and improving stability.

3-69. Engineers will be challenged to understand the operational environment faced and apply the knowledge and background to add to the overall understanding. The engineer view must be consistent with the shared framework and variables employed to analyze the operational environment. But while the levels of conflict and corresponding politically motivated violence may vary in different areas of the world and within a theater, the challenges and opportunities identified by an engineer understanding of the operational environment remain consistently high across the range of military operations. Similarly, the engineer view of the operational environment provides relevant and, sometimes, a unique understanding at each level of war.

OTHER TASKS

3-70. Engineers also participate in or perform a number of other processes that address specific engineer functional requirements or support the integration of engineer activities with the overall operation. Force projection is critical to ensuring that engineer forces are available to execute engineer missions when needed. Engineers plan for the acquisition, construction, management, and disposal of facilities to support the force, and they use project management to complete projects that meet expectations for quality, timeliness, and cost.

FACILITIES AND CONSTRUCTION PLANNING

3-71. Engineers must plan for the acquisition, management, and ultimate disposal of uncontaminated land and facilities, to include—
- Operational facilities (base camps, command posts, airfields, ports).
- Logistics facilities (maintenance facilities, supply points, warehouses, ammunition supply points, waste management areas and facilities, aerial ports of debarkation, seaports of debarkation) for sustainment.
- Force bed-down facilities (dining halls, billeting facilities, religious support facilities, clinics, hygiene facilities).
- Common-use facilities (roads and facilities for joint reception, staging, onward movement, and integration).
- Protection facilities (such as site selection, proximity to potential threat areas, and sniper screening).

3-72. Environmental baseline surveys and occupational environmental health site assessments must be planned before and after site selection or the use of the facilities to ensure minimal exposure to contaminants and to mitigate risks to the U.S. government upon use. (See FM 3-34.5 for additional information.)

3-73. The commander determines which facilities are needed to satisfy operational requirements. Facilities are grouped into six broad categories that emphasize the use of existing assets over new construction. To the maximum extent possible, facilities or real estate requirements should be met from these categories in the following priority:
- U.S.-owned, -occupied, or -leased facilities (including captured facilities).
- U.S.-owned facility substitutes that are pre-positioned in-theater.
- Facilities provided at specified times in designated locations through existing HN and multinational support agreements.
- Facilities available from commercial sources.
- U.S.-owned facility substitutes that are available in the United States.
- Newly constructed facilities that are considered a shortfall after an assessment of the availability of existing assets.

3-74. The engineer staff should plan the expeditious construction of facility requirements that are considered shortfalls (such as those facilities that cannot be sourced from existing assets). In these circumstances, the appropriate Service, HN, alliance, or coalition should perform construction during peacetime to the extent possible. Contracting support should be used to augment military capabilities. If time constraints prevent new construction from being finished in time to meet mission requirements, the engineer should seek alternative solutions to new construction. Expedient construction (such as rapid construction techniques like prefabricated buildings or sand clamshell structures) should also be considered as these methods can be selectively employed with minimum time, cost, and risk.

3-75. Adequate funding must be available to undertake the early engineer reconnaissance and acquisition of facilities to meet requirements, whether by construction or leasing. (See JP 3-34 for additional information.) Funding constraints are a planning consideration. The commander articulates funding requirements for the construction and leasing of facilities by considering the missions supported and the amount of funds required. Funding requirements include facility construction, associated contract administration services, and real estate acquisition and disposal services. Facility construction planning must be accomplished routinely and repetitively to ensure that mission-essential facilities are identified before they are needed and, when possible, that on-the-shelf designs are completed to expedite facility construction.

3-76. The CCDR, in coordination with Service components and the Services, specifies the construction standards for facilities in the theater to optimize the engineer effort expended on any given facility, while assuring that the facilities are adequate for health, safety, and mission accomplishment. The bed-down and basing continuum highlights the need for early master planning efforts to help facilitate the transition to more permanent facilities as an operation develops. While the timelines provide a standard framework, the situation may warrant deviations from them. In addition to using these guidelines when establishing initial construction standards, the Joint Facilities Utilization Board should be used to periodically revalidate construction standards based on current operational issues and provide recommendations to the commander on potential changes. Ultimately, the CCDR determines the exact construction type based on the location, materials available, and other factors. Construction standards are guidelines, and the engineer must consider other planning factors. (See FM 3-34.400 and JP 3-34 for additional discussions of construction standards.)

3-77. Unified facilities criteria provide facility planning, design, construction, operations, and maintenance criteria for DOD components. Individual unified facilities criteria are developed by a single-disciplined working group and published after careful coordination. They are jointly developed and managed by the USACE, the Naval Facilities Engineering Command, and the Air Force Civil Engineer Center. Although unified facilities criteria are written with long-term standards in mind, planners who are executing under contingency and enduring standards for general engineering tasks may find them compulsory. Topics include pavement design, water supply systems, military airfields, concrete design and repair, plumbing, and electrical systems.

3-78. Unified facilities criteria are living documents and will be periodically reviewed, updated, and made available to users as part of the Services' responsibility for providing technical criteria for military construction. Unified facilities criteria are effective upon issuance and are distributed only in electronic media from the following sources:

- Unified Facilities Criteria Index.
- Unified Facilities Criteria Library.
- Naval Facilities Engineering Criteria and Programs Office.
- Construction Criteria Base Index.

3-79. General engineer planners must consider the construction standards established by CCDRs and ASCCs for the area of responsibility. These constantly evolving guidebooks specifically establish base camp standards that consider regional requirements for troop living conditions and, therefore, have a major impact on projects (base camps, utilities). Because the availability of construction materials may vary greatly in various areas of responsibility, standards of construction may differ greatly between them. CCDRs also often establish standards for construction in OPORDs and fragmentary orders that may take precedence over guidebooks. Planners must understand the expected life cycle of a general engineering project to apply these standards. Often the standards will be markedly different, depending on whether the construction is nonpermanent or is intended to be permanent.

Project Management

3-80. Planners use the project management process described in TM 3-34.42 as a tool for coordinating the skill and labor of personnel using equipment and materials to form the desired structure. The project management process (see FM 3-34.400) divides the effort into preliminary planning, detailed planning, and project execution. Today, when engineer planners are focused on general engineering tasks, they often rely on the Theater Construction Management System to produce the products required by the project management system. These products include the design, the activities list, the logic network, the critical path method or Gantt chart, the bill of materials, and other products. Effective products produced during the planning phases also greatly assist during the construction phase. In addition to the Theater Construction Management System, the engineer has various other reachback tools and organizations that can exploit resources, capabilities, and expertise which are not organic to the unit that requires them. Examples of such tools and organizations include the U.S. Army Engineer School, the USACE Reachback Operations Center (412th and 416th TECs), the Air Force Civil Engineer Support Agency, and the Naval Facilities Engineer Command.

3-81. The project management process normally begins at the unit level with the construction directive. This gives the who, what, when, where, and why of a particular project and is similar to an OPORD in its scope and purpose. Critical to the construction directive are plans, specifications, and the items essential for project success. Units may also receive general engineering missions as part of an OPORD, a fragmentary order, a warning order, or verbally. When leaders analyze a construction directive, they may need to treat it as a fragmentary order in that much of the information required for a thorough mission analysis may exist in an OPORD that is issued for a specific contingency operation.

Glossary

SECTION I – ACRONYMS AND ABBREVIATIONS

ADP	Army doctrinal publication
ADRP	Army doctrine reference publication
AJP	allied joint publication
ASCC	Army service component command
ASCOPE	areas, structures, capabilities, organizations, people, and events
ATP	Army techniques publication
attn	attention
ATTP	Army tactics, techniques, and procedures
BCT	brigade combat team
BEB	brigade engineer battalion
CCDR	combatant commander
CCR	U.S. Central Command regulation
DA	Department of the Army
DC	District of Columbia
DOD	Department of Defense
DOTMLPF	doctrine, organization, training, material, leadership and education, personnel, and facilities
DSCA	defense support of civil authorities
EAB	echelons above brigade
EOD	explosive ordnance disposal
FEST	forward engineer support team
FEST-A	forward engineer support team-advance
FEST-M	forward engineer support team-main
FFE	field force engineering
FM	field manual
G-4	assistant chief of staff, logistics
HN	host nation
JFOB	joint forward operations base
JP	joint publication
JTF	joint task force
LOC	line of communication
M/CM/S	mobility, countermobility, and survivability
MEB	maneuver enhancement brigade
MSCoE	U.S. Army Maneuver Support Center of Excellence
NATO	North Atlantic Treaty Organization
No.	number
OAKOC	observation and fields of fire, avenues of approach, key terrain, obstacles, and cover and concealment

Glossary

OPLAN	operation plan
OPORD	operation order
PMESII-PT	political, military, economic, social, information, infrastructure, physical, environment, and time
S-4	logistics staff officer
SWEAT-MSO	sewage, water, electricity, academics, trash medical, safety, and other considerations
TEC	theater engineer command
U.S.	United States
USACE	U.S. Army Corps of Engineers
USC	U.S. Code

SECTION II – TERMS

*countermobility operations
   Those combined arms activities that use or enhance the effects of natural and man-made obstacles to deny an adversary freedom of movement and maneuver.

*engineer work line
   A coordinated boundary or phase line used to compartmentalize an area of operations to indicate where specific engineer units have primary responsibility for the engineer effort.

*field force engineering
   The application of the Engineer Regiment capabilities from the three engineer disciplines (primarily general engineering) to support operations through reachback and forward presence.

References

REQUIRED PUBLICATIONS
These documents must be available to the intended users of this publication.
ADRP 1-02. *Terms and Military Symbols.* 24 September 2013.
JP 1-02. *Department of Defense Dictionary of Military and Associated Terms.* 8 November 2010.

RELATED PUBLICATIONS
These documents contain relevant supplemental information.

ARMY
Most Army doctrinal publications are available online at <www.apd.army.mil>.
ADP 2-0. *Intelligence.* 31 August 2012.
ADP 3-0. *Unified Land Operations.* 10 October 2011.
ADP 3-07. *Stability.* 31 August 2012.
ADP 3-28. *Defense Support of Civil Authorities.* 26 July 2012.
ADP 3-37. *Protection.* 31 August 2012.
ADP 3-90. *Offense and Defense.* 31 August 2012.
ADP 4-0. *Sustainment.* 31 July 2012.
ADP 5-0. *The Operations Process.* 17 May 2012.
ADP 6-0. *Mission Command.* 17 May 2012.
ADP 6-22. *Army Leadership.* 1 August 2012.
ADRP 3-0. *Unified Land Operations.* 16 May 2012.
ADRP 3-07. *Stability.* 31 August 2012.
ADRP 3-28. *Defense Support of Civil Authorities.* 14 June 2013.
ADRP 3-37. *Protection.* 31 August 2012.
ADRP 5-0. *The Operations Process.* 17 May 2012.
ADRP 6-0. *Mission Command.* 17 May 2012.
ATP 3-37.10. *Base Camps.* 26 April 2013.
ATP 3-37.34. *Survivability Operations.* 28 June 2013.
ATP 4-94. *Theater Sustainment Command.* 28 June 2013.
ATTP 3-34.23. *Engineer Operations–Echelons Above Brigade Combat Team.* 8 July 2010.
ATTP 3-34.80. *Geospatial Engineering.* 29 July 2010.
ATTP 3-34.84. *Multi-Service Tactics, Techniques, and Procedures for Military Diving Operations.* 12 January 2011.
ATTP 3-90.4. *Combined Arms Mobility Operations.* 10 August 2011.
ATTP 5-0.1. *Commander and Staff Officer Guide.* 14 September 2011.
FM 3-06. *Urban Operations.* 26 October 2006.
FM 3-07. *Stability Operations.* 6 October 2008.
FM 3-16. *The Army in Multinational Operations.* 20 May 2010.
FM 3-34.5. *Environmental Considerations.* 16 February 2010.
FM 3-34.22. *Engineer Operations–Brigade Combat Team and Below.* 11 February 2009.
FM 3-34.170. *Engineer Reconnaissance.* 25 March 2008.
FM 3-34.210. *Explosive Hazards Operations.* 27 March 2007.

References

FM 3-34.400. *General Engineering.* 9 December 2008.
FM 3-90.31. *Maneuver Enhancement Brigade Operations.* 26 February 2009.
FM 4-92. *Contracting Support Brigade.* 12 February 2010.
FM 5-415. *Fire-Fighting Operations.* 9 February 1999.
FM 27-10. *The Law of Land Warfare.* 18 July 1956.
FM 90-7. *Combined Arms Obstacle Integration.* 29 September 1994.
GTA 05-08-002. *Environmental-Related Risk Assessment.* 1 March 2008.
TC 2-22.7. *Geospatial Intelligence Handbook.* 18 February 2011.
TM 3-34.42. *Construction Project Management.* 1 December 2010, available at <https://ndls.nwdc.navy.mil/>, accessed on 7 January 2014.
TM 3-34.83. *Engineer Diving Operations.* 2 August 2013.

JOINT

Most joint publications are available online at <www.dtic.mil/doctrine/new_pubs/jointpub.htm>.
GTA 90-01-011. *Joint Forward Operations Base (JFOB) Protection Handbook.* October 2011.
JP 1. *Doctrine for the Armed Forces of the United States.* 25 March 2013.
JP 2-03. *Geospatial Intelligence in Joint Operations.* 31 October 2012.
JP 3-15. *Barriers, Obstacles, and Mine Warfare for Joint Operations.* 17 June 2011.
JP 3-34. *Joint Engineer Operations.* 30 June 2011.

NATO

AJP 3.12(A). *Allied Doctrine for Military Engineer Support to Joint Operations.* 28 September 2010.

OTHER SOURCES

10 USC 2675. *Leases: Foreign Countries.* 7 January 2011.
CCR 415-1. *Construction and Base Camp Development* (commonly known as *The Sand Book*). 15 April 2009.
Doctrine 2015 Strategy. June 2011.
Emergency Support Function 3–Public Works and Engineering Course, available at <http://training.fema.gov/EMIWeb/IS/courseOverview.aspx?code=is-803>, accessed on 11 December 2013.
Engineer Pamphlet 500-1-2. *Field Force Engineering–United States Army Corps of Engineers Support to Full Spectrum Operations.* 1 August 2010.
National Support Framework, May 2013, available at <http://www.fema.gov/library/viewRecord.do?id=7371>, accessed on 11 December 2013.
U.S. European Commander's Camp Facilities Standards for Contingency Operations (commonly known as *The Red Book*). 1 February 2004.

PRESCRIBED FORMS

None.

REFERENCED FORMS

Most Army forms are available online at <www.apd.army.mil>.
DA Form 2028. *Recommended Changes to Publications and Blank Forms.*

WEB SITES

Army Knowledge Online, Doctrine and Training Publications Web site, <https://armypubs.us.army.mil/doctrine/index.html>, accessed on 10 December 2013.

References

Army Publishing Directorate, Army Publishing Updates Web site, <http://www.apd.army.mil/AdminPubs/new_subscribe.asp>, accessed on 10 December 2013.

Construction Criteria Base Index, National Institute of Building Sciences Web site, <http://www.wbdg.org/ccb/ccb.php>, accessed on 11 December 2013.

Naval Facilities Engineering Criteria and Programs Office, National Institute of Building Sciences Web site, <http://www.wbdg.org/references/pa_dod_cieng.php>, accessed on 11 December 2013.

Unified Facilities Criteria Index, National Institute of Building Sciences Web site, <http://www.wbdg.org/ccb/browse_cat.php?c=4>, accessed on 11 December 2013.

Unified Facilities Criteria Library, USACE Protective Design Center Web site, <https://pdc.usace.army.mil/library/ufc/>, accessed on 11 December 2013.

USACE Reachback Operations Center Web site, <https://uroc.usace.army.mil>, accessed on 11 December 2013.

This page intentionally left blank.

Index

Entries are by paragraph number.

A

Army forces, 3-13
assured mobility, 2-2, 2-22, 3-5

C

close combat, 1-10, 2-2, 2-3, 2-5, 2-7, 2-9, 2-11
combat engineer units, 1-10, 2-2, 2-5, 2-15
combat engineering, 1-1, 1-10, 1-19, 2-2, 2-3, 2-5, 2-9, 2-11, 2-12, 2-15, 2-17, 2-21, 3-4
    definition, 1-1
combat engineers, 1-1, 1-12, 2-3, 2-6, 2-11, 2-14, 2-15
combat power, iii, 1-1, 2-5, 2-8, 2-9, 2-10, 2-14
combined arms, 1-1, 1-4, 1-10, 2-1, 2-3, 2-9, 2-15, 2-17, 2-19, 2-21, 3-1, 3-6, 3-9, 3-18
countermobility, 2-3, 2-4, 2-6, 2-16, 2-17, 3-5, 3-13

E

engineer disciplines, 1-2, 1-3, 2-1, 2-5, 2-6, 2-9, 2-11, 2-15
engineer operations, 1-18, 2-10, 3-8, 3-9, 3-11
engineer reconnaissance teams, 1-12, 2-13
Engineer Regiment, iv, 2-6, 2-19, 2-20, 3-5
engineer work line, 2-10
    definition, 2-10

F

field force engineering (FFE), 2-8, 2-14

force projection, 1-20, 1-21, 2-1, 2-2, 2-6, 3-2, 3-18
forward engineer support team (FEST), 3-3, 3-14

G

general engineer, 2-5, 2-12, 2-14, 2-16, 2-17, 2-18, 2-19, 2-20, 3-20
general engineer unit, 1-10, 1-11, 2-2, 2-3, 2-5
general engineering, 1-2, 1-10, 1-12, 2-2, 2-6, 2-9, 2-10, 2-14, 2-19, 2-22, 3-17, 3-20
    definition, 1-2
geospatial engineering, iv, 1-4, 1-11, 2-1, 2-2, 2-4, 2-9, 2-11, 2-15, 2-17, 2-18, 3-6, 3-8, 3-17, 3-18
geospatial information and services, 2-13
    definition, 2-13

L

lethal, 2-4, 2-9, 2-10
levels of war, 2-2, 2-5, 3-1, 3-2, 3-14
lines of engineer support, v, vi, 2-1, 2-4, 3-7
    definition, 2-1

M

maneuver, 1-10, 2-1, 2-2, 2-3, 2-5, 2-7, 2-9, 2-11, 2-14, 2-15, 2-16, 2-17, 2-19, 2-21, 2-22, 3-1, 3-4, 3-6, 3-9, 3-11, 3-12, 3-13, 3-16, 3-17, 3-18
mission command, 1-9, 1-10, 1-11, 1-20, 2-8, 2-9, 2-10, 2-16, 2-20

movement and maneuver warfighting function, 2-3
    definition, 2-11
multinational, 1-9, 1-11, 1-20, 1-21, 3-7

N

nonlethal, 2-4, 2-9, 2-10, 3-16

O

obstacles, 1-10, 2-1, 2-3, 2-5, 2-6, 2-14, 2-15, 2-16, 2-17, 3-7, 3-8, 3-9, 3-11
operational environment, 1-12, 3-7, 3-8, 3-18

P

protection, 2-5
protection warfighting function, 2-4, 2-9, 2-14, 2-18, 3-4
    defninition, 2-14

S

survivability operations, 2-5, 2-11, 2-14
sustainment warfighting function, 2-12, 2-14, 2-16, 3-4, 3-17
    definition, 2-14

T

task-organized, 1-10, 2-15, 2-17, 3-5, 3-7, 3-12, 3-17

U

unified action, 3-8
unified land operations, iii, 1-12, 2-1, 2-4, 2-5, 2-8, 2-9, 2-11, 2-14, 2-15, 2-19, 3-16, 3-17, 3-18
    definition, 3-17

This page intentionally left blank.

FM 3-34
2 April 2014

By order of the Secretary of the Army:

RAYMOND T. ODIERNO
General, United States Army
Chief of Staff

Official:

GERALD B. O'KEEFE
Administrative Assistant to the
Secretary of the Army
1402805

DISTRIBUTION:
Active Army, Army National Guard, and U.S. Army Reserve: To be distributed in accordance with the initial distribution number (IDN) 110451, requirements for FM 3-34.

This page intentionally left blank.

www.ingramcontent.com/pod-product-compliance
Lightning Source LLC
Chambersburg PA
CBHW071756170526
45167CB00003B/1058